SCIENCE MATTERS

The Rise and Fall of Leaded Petrol

prepared for the Course Team by David Johnson

Science: a second level course

The S280 Course Team

Pam Berry (Text Processing)
Norman Cohen (Author)
Angela Colling (Author)
Michael Gillman (Author)
John Greenwood (Librarian)
Barbara Hodgson (Reader)
David Johnson (Author)
Carol Johnstone (Course Secretary)
Hilary MacQueen (Author)
Isla McTaggart (Course Manager)
Diane Mole (Designer)
Joanna Munnelly (Editor)
Pat Murphy (Author)
Ian Nuttall (Editor)
Pam Owen (Graphic Artist)
Malcolm Scott (Author)
Sandy Smith (Author)
Margaret Swithenby (Editor)
Jeff Thomas (Course Team Chair and Author)
Kiki Warr (Author)
Bill Young (BBC Producer)
External Assessor: John Durant

The Open University, Walton Hall, Milton Keynes, MK7 6AA.

First published 1993. Reprinted 1996.

Edited, designed and typeset in the United Kingdom by the Open University.

Printed in the United Kingdom by Thanet Press Ltd, Margate, Kent.

ISBN 07492 51018

This text forms part of an Open University Second Level Course. If you would like a copy of *Studying with the Open University*, please write to the Central Enquiry Service, PO Box 200, The Open University, Walton Hall, Milton Keynes, MK7 6YZ. If you have not already enrolled on the Course and would like to buy this or other Open University material, please write to Open University Educational Enterprises Ltd, 12 Cofferidge Close, Stony Stratford, Milton Keynes, MK11 1BY, United Kingdom.

1.2

4895C/s280lipi1.2

Contents

1 Two faces of science

At about the time that the pens of the authors of *Science Matters* made their first reluctant contact with paper, Dr Jonathan Miller took part in a Channel Four TV programme that involved three guests and a host. Each guest in turn introduced a media item that had taken their fancy in the preceding week, and then participated in a discussion of it with the others. When Miller's turn came, he chose extracts from two Open University TV programmes. One had been produced by the Science Faculty, and showed Dr Bob Lambourne discussing the central role of the Schrödinger equation in quantum mechanics; the other was an Arts Faculty programme about the impact of the monastery on medieval society. Miller went on to say that he believed that when people later looked back, they would set the Open University alongside the National Health Service as one of the two great socialist achievements of 20th century Britain. Whether you regard this as an insult or a compliment will depend upon your politics.

Miller then went on to give reasons for his choices. The two programmes represented two distinct kinds of human interest that affect the way in which we think about the world. From one standpoint, our world seems remote and untouchable: the behaviour of stars, planets, chemicals, molecules, atoms and electrons, is governed by rules that we seem unable to influence. But from this realization comes an impulse to understand that behaviour by formulating general laws which describe it, and which, at their best, can predict it in advance. The fruits of that impulse are the laws of physics and chemistry, and if we judge them by their sweep and fertility, the Schrödinger equation is one of the supreme achievements: it can, in principle, predict the outcome of any chemical experiment. Fortunately, there are great mathematical difficulties in solving the equation in all but the simplest cases, so research chemists can still find employment; this, however, in no way diminishes the standing of the equation among the laws of natural science.

Figure 1.1 Erwin Schrödinger (1887–1961). He was born in Austria, but with British ancestry on his mother's side and spoke almost flawless English. He graduated from the University of Vienna where he also did research in theoretical physics before serving as an artillery officer in the Austro-Hungarian army during the First World War. His most creative work, including his famous equation, was done at Zurich where he was Professor from 1921 to 1927. A move to Berlin in 1927 signalled the beginning of 12 restless years during which he struggled to distance himself from the emerging Nazi regime. He moved to Oxford, then Graz in Austria, but was forced to leave after the Nazi Anschluss of 1938. After spending time in Rome, Geneva, Oxford and Ghent, in late 1939 Schrödinger found refuge at the Dublin Institute of Advanced Study where he was appointed founding Professor by the Irish President, Eamonn de Valera, who was also an accomplished mathematician. Schrödinger worked in Dublin for 17 years, before returning to Austria, where he died in 1961.

Now although the laws of nature may be unalterable, the world which they describe is not. And that brings us to the second of Miller's two kinds of human interest: an impulse to understand the way in which our environment, both natural and social, is changed by us. This underlies the study of archaeology, history, sociology and economics, and is clearly represented by the programme on the relations that existed between the medieval monastery and the surrounding society.

What Miller did not say was that the division that he identified exists within science itself. Traditionally, science has been taught out of the first kind of interest: the search for the laws of nature. This is the way of the Science Foundation Course which describes how that search was served by figures such as Newton, Einstein, Heisenberg, Schrödinger, Mendeléev, Darwin, Mendel, Wegener and others. But the priorities of these people are not shared by all scientists. They did not, for example, regard the practical applications of their work as of paramount importance, and in some cases, their temperaments seemed ill-fitted for a collaborative world like industry. Here, for instance, are some remarks about Schrödinger (Figure 1.1) by a fellow scientist who worked with him at the Dublin Institute of Advanced Study:

> *The Institute was very suitable and comfortable for quiet work unburdened by mass lecturing and the modern curse of the scientists—administration... Throughout his life, Schrödinger was a lone worker. Very rarely, he accepted research students. There were a few, though, in Dublin. There is hardly any publication with joint authorship of which Schrödinger was a partner. Nothing*

Figure 1.2 Paul Dirac (1902–1984). Dirac was born in Bristol, and took degrees in electrical engineering and mathematics at Bristol University before moving to Cambridge in 1923. The paper in which he made the Schrödinger equation compatible with the theory of relativity was described by Sir Neville Mott, himself a Nobel Prize winner, as 'the most beautiful and exciting piece of pure theoretical physics I have seen in my lifetime'. One field in which Dirac did not excel was driving: it was once remarked that the energy of his car was quantized with just two allowed speeds: zero and flat out.

could have been more foreign to him than the now fashionable 'team work'. All his work bore the imprint of personality and individualism.

Schrödinger's temperament was very similar to that of the British physicist, Paul Dirac, with whom he shared the Nobel Prize for Physics in 1933 (Figure 1.2). Dirac discovered how the Schrödinger equation could be made compatible with the theory of relativity. His unworldliness and love of privacy are revealed by the exchange that took place when Rutherford arrived to offer his congratulations. Dirac said that he would refuse the Nobel Prize because he could not bear all the publicity. This embarrassing prospect was only avoided when Rutherford pointed out that a refusal would attract even more publicity: Dirac then changed his mind, and went to the ceremony in Stockholm after all. Some years later, during a lecture at Moscow University, Dirac was asked for his philosophy of science; he wrote on a blackboard, 'Physical laws should have mathematical beauty', and to this day, the blackboard, with his writing, has been preserved in Moscow.

What is so obviously under-represented, when we read of Schrödinger and Dirac, are the very things for which ordinary people value science: its usefulness, and its power to achieve intended effects. Nor do we get a sense of the fear that science may inspire through its capacity to achieve *unintended* effects as well. Through such powers, science has changed our society and our environment, and their influence obviously falls into the second of the two categories of interest defined by Jonathan Miller: the study of the ways in which we have altered our world. Even Dirac could not stay permanently aloof from this other side of science. During the Second World War, when all combatants found themselves desperately short of scientists, he was briefly drawn into the project to manufacture an atomic bomb. Today, the most economical method of enriching uranium for use in nuclear reactors and nuclear weapons, employs a centrifuge designed to separate gases, and the fundamental mathematical equations which govern the centrifugal action were first worked out by Dirac in 1941.

Science Matters represents this other side of science, its impact upon our world. It complements the Science Foundation Course, where the emphasis is on fundamental theories and principles. In *Science Matters* the stress is on applying ideas that you have already encountered, in new situations, and there will be revision and reminders of what you need to know in order to do this.

The Course consists of a series of books on quite distinct topics. Each topic is currently arousing, or has recently aroused, widespread public interest or concern. Every book is self-contained, and so constitutes a fresh start: it can be read and understood even if you have read none of the others. We hope that because of this feature your studies will not be as relentless as for other Courses.

1.1 Skills in Science Matters

As you study *Science Matters*, you will realize not just that its philosophy is different from that of the Science Foundation Course; you will also find yourself carrying out new and different tasks. For example, you will extract information from published scientific articles, and use it to formulate your own views in your own words; you will need to recognize the incompleteness or unreliability of some scientific data, and to make judgements about the limitations of science in addressing social problems. As you work through them, exercises of this sort will provide you with new kinds of expertise to set alongside those provided by the Science Foundation Course. At the end of each book, the exercises are summarized as a list of skills, and against each skill are references to the activities and/or questions in which it is practised. We hope that this explicit description of what is expected of you, will be helpful.

Alongside the new skills which you meet in *Science Matters*, you will use others that you acquired in the Science Foundation Course. In this category come essential numerical skills, and one of them, the manipulation of quantities in scientific notation, is revised now. This is because its importance is such that it will be needed in the next chapter, and in most of the other books in the Course. To see whether you need to revise this topic, you should attempt Questions 1.1 and 1.2 below. If you have difficulty with them read Box 1.1, which reminds you of the technique.

Question 1.1 The mass of the electron is 9.110×10^{-31} kg. Numbers which are written in such a form—an ordinary decimal multiplied by ten raised to some particular power—are said to be in scientific notation. To check that you understand, and can manipulate quantities of this type, try the following questions:

(a) Write the quantities (i) 420 000 000 kg and (ii) 0.000 000 23 m in scientific notation.

(b) Now reverse the process, and write the quantities (i) 3.6×10^{5} kg and (ii) 8.5×10^{-5} m using ordinary decimal numbers.

Question 1.2 Without using a calculator, perform and give answers to the following calculations:

(a) $(4.2 \times 10^{8}\,\text{kg}) \times (2.0 \times 10^{6}\,\text{m}\,\text{s}^{-1})$

(b) $(4.2 \times 10^{8}\,\text{m}) \div (2.0 \times 10^{6}\,\text{s})$

(c) $(4.2 \times 10^{6}\,\text{m}\,\text{s}^{-1}) \times (2.0 \times 10^{-4}\,\text{s})$

(d) $(4.2 \times 10^{3}\,\text{m}\,\text{s}^{-1}) \div (2.0 \times 10^{-4}\,\text{s})$

1.1

Box 1.1 Quantities in scientific notation

Science is often concerned with quantities which are very much larger or smaller than those of everyday experience. For example, an adult woman may have a height of 1.54 m. By contrast, the Sun is typically about 154 000 000 000 m from the Earth, and the atoms of carbon in diamond are separated by only 0.000 000 000 154 m. There is a very much more concise way of writing out such very large and very small quantities.

Let us start with our everyday distance of 1.54 m. We can multiply it by 10 by advancing the decimal point by one digit to give 15.4 m. Multiplication by 100 requires an advance of two digits to 154 m. If we go one step further and multiply by 1 000, we must advance the decimal point by three digits, but our initial distance of 1.54 m has only two digits for the decimal point to jump. We provide the extra one by adding a zero to the end of the number, and the three jumps then give 1 540 m. Thus 1 540 m is the same as $(1.54 \times 1\,000)$ m. Now 1 000 is more concisely written as 10^{3} where the power three is the number of zeros after the one. So 1 540 m can be written 1.54×10^{3} m. This way of writing a number, as an ordinary decimal multiplied by a power of ten, is often called **scientific notation**. In scientific notation, it is conventional to arrange things so that the decimal lies between 1 and 10. For example 1 540 m could also be written correctly as 0.154×10^{4} m, but 1.54×10^{3} m is more conventional.

▷ Write down the distance from the Earth to the Sun in scientific notation.

▶ 1.54×10^{11} m; if you start with 1.54 m the decimal point must be advanced 11 places to give 154 000 000 000 m.

Suppose now that we *divide* 1.54 m by 1 000: the decimal point now *retreats* three places, and, inserting the necessary zeros before the number for the decimal point to jump, we obtain 0.001 54 m. This is written as 1.54×10^{-3} m in scientific notation, where the power −3 marks the three-digit retreat of the decimal point, in contrast to the three-digit advance called for with the distance 1.54×10^{3} m.

▷ Write down the distance between the carbon atoms in a diamond in scientific notation.

▶ 1.54×10^{-10} m; if you start with 1.54 m, the decimal point must retreat ten places to give 0.000 000 000 154 m.

Notice that 1.54×10^{-10}, the multiplication of a decimal by 10^{-10}, means the division of that decimal by 10^{10}.

Numbers in scientific notation can be added or subtracted by first converting them back into ordinary decimal form. However, in this, and subsequent books in the Course, you will be concerned mainly with multiplication and division, and it is on these operations that we concentrate here. A number in scientific notation has two parts: an ordinary decimal (e.g. 1.54) and a power of ten (e.g. 10^{-11}). When two such numbers are multiplied or divided, the two parts of each number are multiplied or divided separately.

For example, suppose a rectangle has sides of 3×10^3 m and 2×10^4 m. Its area is obtained by multiplying the lengths of the sides together. Multiplying the ordinary decimal parts gives us $3 \times 2 = 6$. Multiplying the powers of ten gives us $10^3 \times 10^4 = 10^7$, because when powers of ten are multiplied, one simply adds the powers. Thus,

$$\text{Area} = (3 \times 10^3\,\text{m}) \times (2 \times 10^4\,\text{m})$$
$$= 6 \times 10^7\,\text{m}^2$$

Notice that it is not just the numbers that are multiplied, the units get the same treatment: $\text{m} \times \text{m} = \text{m}^2$.

To illustrate the process of division, we start with a light-year. This is the distance a ray of light travels in one year, and is equal to 9.46×10^{15} m. Now a year, expressed in seconds, is 3.15×10^7 s. Thus the speed of light, c, is given by the distance travelled divided by the time:

$$c = \frac{9.46 \times 10^{15}\,\text{m}}{3.15 \times 10^7\,\text{s}}$$

Treating the decimal, the power parts and the units separately:

$$c = \frac{9.46}{3.15} \times \frac{10^{15}}{10^7} \times \frac{\text{m}}{\text{s}}$$
$$= 3.00 \times \frac{10^{15}}{10^7} \times \frac{\text{m}}{\text{s}}$$

Remembering that when powers of ten are divided, the power superscripts are subtracted:

$$c = 3.00 \times 10^8 \times \frac{\text{m}}{\text{s}}$$
$$= 3.00 \times 10^8\,\text{m}\,\text{s}^{-1}$$

Again, the symbols for the units are multiplied or divided like the numbers.

Usually you will manipulate numbers in scientific notation with a calculator; this box should have shown you how you could dispense with it if need be. ■

1.2 Choosing an introductory topic

You now know, in a general sense, how *Science Matters* differs from the Science Foundation Course: there is an emphasis on skills, and on topics of current public interest, topics that illustrate the way in which science affects the world we live in. What subject then provides a suitable beginning? It is helpful to choose something that continues to get considerable media attention, but which received its maximum dose of publicity in the recent past. In the next section, we give reasons for this, and also for the particular choice that the Course Team has made: *The Rise and Fall of Leaded Petrol.*

1.3 Introducing The Rise and Fall of Leaded Petrol

On Monday 18 April 1983, Tom King, the then Environment Secretary, told the House of Commons that by 1990, all new cars sold in Britain would have to run on lead-free fuel. His decision had been prompted by the 9th Report of the Royal Commission on Environmental Pollution, subtitled *Lead in the Environment*. It was only a little over 60 years since the day—1 February 1923—when the first litre of lead-improved petrol had been sold to an adventurous motorist in Dayton, Ohio (Figure 1.3); you will learn why in Section 3.3.4. For the British, Tom King's announcement

Figure 1.3 The garage in Dayton, Ohio where, at about 10 a.m. on the morning of 1 February 1923, the first gallon of leaded petrol was sold. The dispensing pump is on a concrete block, and is fitted with a metering device so the customers can see the petrol and liquid lead compound mixed before their own eyes.

signalled the end for a practice which had once been described as 'one of the most significant contributions to oil and automotive progress'.

The environmental lobby basked in the unfamiliar sensation of almost total victory. Writing in *The Times* the following day, Des Wilson, the chairman of CLEAR, the Campaign for LEad-free AiR, combined triumphalism with suspicion. It was important that 'Government and the multinational industries should learn the crucial lessons' from the controversy. His organization's success ensured that from now on, environmental issues would be 'placed higher on the political agenda', and the public had shown that when the evidence of risk from pollution is substantial, 'they expect the necessary action, and will pay the price'. At the same time, the target date of 1990 was too remote. Although CLEAR did not have 'the benefit of its opponents' technical and propaganda resources', it was sure that the multinationals had exaggerated the costs and difficulties of the move to lead-free petrol. Especially reprehensible was the Royal Commission's recommendation that government should work with just the oil industries and car manufacturers to establish a timetable for the change. Why could not organizations such as CLEAR, whose endeavours had led to the Government's decision, also be included? Scepticism remained about the determination of ministers 'to act with resolution'.

Yet any disinterested student of the controversy could see that the environmental lobbies had cause for self-congratulation. Painstaking but carefully selected science, statistical methods, presentational skills and political activity had all been astutely combined to work the issue of lead pollution into the papers, and on to radio and television. From this flowed the widespread public and scientific concern that had persuaded the Royal Commission on Environmental Pollution to devote its 9th Report specifically to lead, rather than to more general issues. From a political standpoint, that report had, in its turn, settled the outcome of the controversy over lead in petrol—the environmentalists had succeeded and, in the UK, leaded petrol was ultimately doomed.

Yet when military historians discuss the outcome of battles, they look beyond the zeal of troops or the brilliance of generals; they study, for example, the terrain over which the battle was fought, and the weapons that each side deployed. Was there something special about 'lead in petrol' that made the going easy for environmentalists? As you will see in this book, your understanding of basic science will help you to respond to that question, and in so responding, that understanding will be reinforced. This symbiotic process is one that *Science Matters* is designed to encourage, and the subject of lead in petrol serves it well. Another attractive feature is that the topic is a mature scientific issue of public concern that can be examined from its beginnings through to a resolved *political* outcome. As such, it can provide interesting insights into issues studied later in the Course that remain politically unresolved.

To begin with, however, you need to know something about lead, and the circumstances under which it becomes a danger to people. This means, in turn, that you need to be familiar with elementary chemical concepts such as chemical element, chemical formula and chemical equation. Box 1.2 will refresh your memory of these basic concepts, and it is followed by two questions which test your understanding of them.

1.2

Box 1.2 Chemical elements, chemical formulae and chemical equations

Everything that you can see around you is made up of tiny particles called atoms. There are about 100 different types of atom, and each different type is given a name and a symbol. For example, one type is called sulphur, symbol S, another lead, symbol Pb, and another oxygen, symbol O. These symbols for most of the types of atom that you will meet in this book are shown in Table 1.1.

Table 1.1 Symbols for most of the different types of atom (chemical elements) encountered in this book.

Element	Symbol
aluminium	Al
bromine	Br
carbon	C
chlorine	Cl
chromium	Cr
hydrogen	H
lead	Pb
magnesium	Mg
neon	Ne
nitrogen	N
oxygen	O
platinum	Pt
rhenium	Re
rhodium	Rh
silicon	Si
sulphur	S
tellurium	Te

Some materials contain just one type of atom. Both these materials, and the types of atom that they contain, are called **chemical elements.** For example, metallic lead is a chemical element which contains just lead atoms, and is written Pb(s), the 's' in brackets indicating that the material is a solid.

Most substances, however, consist of two or more chemical elements whose atoms are intermingled and bound together in some kind of regular pattern. These substances are called **chemical compounds**. Their compositions can be represented by an **empirical formula** which tells us the proportions in which the different types of atom are combined. Thus the chemical elements lead and oxygen can combine to form compounds called lead oxides. One oxide is an orange powder with the empirical formula PbO; another is a brown powder with the empirical formula PbO_2. Each formula carries two types of information: first, the symbols of the different chemical elements which are present in the compound; and second, subscript numbers which follow those symbols. Where a symbol lacks a following subscript number, that subscript is taken to be one. These numbers tell us the ratios in which the different types of atom are combined in the compound. Thus the empirical formula PbO_2 tells us that in the brown powder there are two oxygen atoms present for every lead atom.

▷ What does the formula PbO tell us about the orange powder?

▶ The (implicit) subscript numbers against Pb and O are both one. There is one oxygen atom present for every lead atom—the orange powder contains equal numbers of lead and oxygen atoms.

In certain cases chemists avoid using empirical formulae, and instead employ formulae that carry more information than just the proportions in which the elements are combined. A very simple, but important case is the gas oxygen which consists of just oxygen atoms. The empirical formula would be O(g), the 'g' in brackets indicating that the substance is a gas. But it is known that oxygen gas consists of pairs of oxygen atoms bound together, and that these pairs move about almost independently of each other. Oxygen gas is therefore written O_2(g). These tiny unit pairs of which the gas is composed are called molecules, and the type of formula is called the **molecular formula**.

Many chemical substances dissolve in water, and when so dissolved, they are said to be in aqueous solution. One very important type of aqueous solution contains charged particles called **ions**. For example, when the gas hydrogen chloride, HCl, dissolves in water, each electrically neutral HCl molecule breaks up into a hydrogen ion, H^+(aq), carrying a single positive charge, and a chloride ion, Cl^-(aq), carrying an equal but opposite, negative, charge. The 'aq' in brackets indicates that the ions are in aqueous solution. The positive hydrogen ion, H^+(aq), is very important because solutions which contain it are **acids**. Thus the solution of HCl in water is called hydrochloric acid.

The ions H^+(aq) and Cl^-(aq) consist of single atoms with just one extra positive and one extra negative charge respectively. Some ions consist of single atoms with more than one extra positive or negative charge. Others consist of collections of atoms bound together, and they, too, may carry more than one unit of positive or negative charge. Aqueous ions that you will meet in this book are listed in Table 1.2.

What atoms do lead ions and sulphate ions contain, and what charges do these ions carry?

Table 1.2 Chemical formulae of some of the aqueous ions encountered in this book.

Name	Formula
acetate	$CH_3CO_2^-$*
carbonate	CO_3^{2-}
chloride	Cl^-
chromate	CrO_4^{2-}
hydrogen	H^+
hydroxide	OH^-
lead	Pb^{2+}
nitrate	NO_3^-
sulphate	SO_4^{2-}

*The reason why the two carbon atoms in the acetate ion are written separately, rather than being grouped together as C_2, will become clear after reading Box 3.1.

▶ Each lead ion contains just one lead atom and carries two positive charges; each sulphate ion contains one sulphur and four oxygen atoms bound together, and carries two negative charges.

The chemical formulae of neutral compounds and ions give meaning to chemical changes. Lead sulphide or *galena*, PbS, is the commonest naturally-occurring form of lead; it occurs widely in parts of Derbyshire and has a shiny blue–black appearance. If heated in air, it reacts with oxygen, O_2(g), to form an orange powder, PbO(s), and poisonous fumes of sulphur dioxide, SO_2(g), which stream off into the atmosphere. The first example in Figure 1.4 shows a **chemical equation** for this change.

The reactants, PbS(s) and O_2(g), appear on the left; the products, PbO(s) and SO_2(g), on the right. As in all chemical changes, individual atoms are not destroyed, but merely rearranged: before the change, the sulphur atoms were bound to lead atoms in

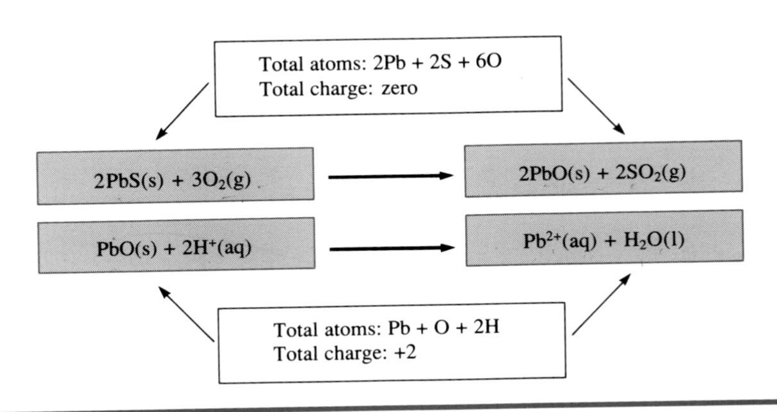

Figure 1.4 Two balanced equations for chemical changes which are described in the text. Each chemical formula is preceded, and therefore multiplied, by a whole number. Where no number precedes a formula, that number is understood to be one. The combination of chosen whole numbers must be such that on both sides of the equation there are equal numbers of the different types of atoms and the same total charge. By convention, the smallest possible combination of whole numbers is chosen.

lead sulphide; afterwards both they and the lead atoms are bound to oxygen. This preservation of the atoms is acknowledged by making the chemical equation a **balanced equation**: each formula is multiplied by a whole number which precedes it in the equation, and the combined effect is to produce the same numbers of each type of atom on both sides of the equation. By convention, the smallest possible combination of whole numbers is used.

Balanced equations must also be balanced with respect to charge: the total charge must be the same on both sides of the equation. In the first example in Figure 1.4, this is automatically guaranteed by the fact that all reactants and products are electrically neutral: the total charge on each side is zero. Not so in the second example, which contains charged ions. It is concerned with what happens when the orange powder PbO(s) which was produced in the first reaction is dropped into acidified water. The lead oxide dissolves, reacting with the hydrogen ions to produce lead ions, Pb^{2+}(aq), and liquid water H_2O(l). ■

The two questions below test your understanding of Box 1.2.

Question 1.3

(a) The empirical formula of solid lead sulphate is $PbSO_4$. What does this tell us about the relative numbers of the different types of atom present in the compound?

(b) The molecular formula of the gas, carbon dioxide, is CO_2. The empirical formula of the solid, lead dioxide, is PbO_2. Which formula gives us the most information, and why?

Question 1.4 If orange lead oxide, PbO(s), is heated on a charcoal fire at about 350 °C, the lead oxide reacts with carbon, C(s), producing molten lead, Pb(l) and carbon dioxide, CO_2(g). Write a balanced equation for the reaction.

2 Lead: its chemistry, toxicity and measurement

Metallic lead has been with us for a long time: in the ruins of Catal Huyuk in southern Turkey, lead beads have been found, and dated to 6 500 BC. Metallic lead hardly ever occurs naturally, and is almost invariably obtained from lead minerals such as galena, so this date suggests that lead was the first metal to be smelted, that is, to be chemically extracted from its ores. As archaeologists have confirmed by trying it, this could have happened accidentally if galena was mistakenly put on a charcoal or wood fire:

$$2PbS(s) + 3O_2(g) \longrightarrow 2PbO(s) + 2SO_2(g) \quad (2.1)$$

$$2PbO(s) + C(s) \longrightarrow 2Pb(l) + CO_2(g) \quad (2.2)$$

Even today, much lead is still obtained by these two reactions, but the amounts are much greater than in earlier times. Figure 2.1 shows how much.

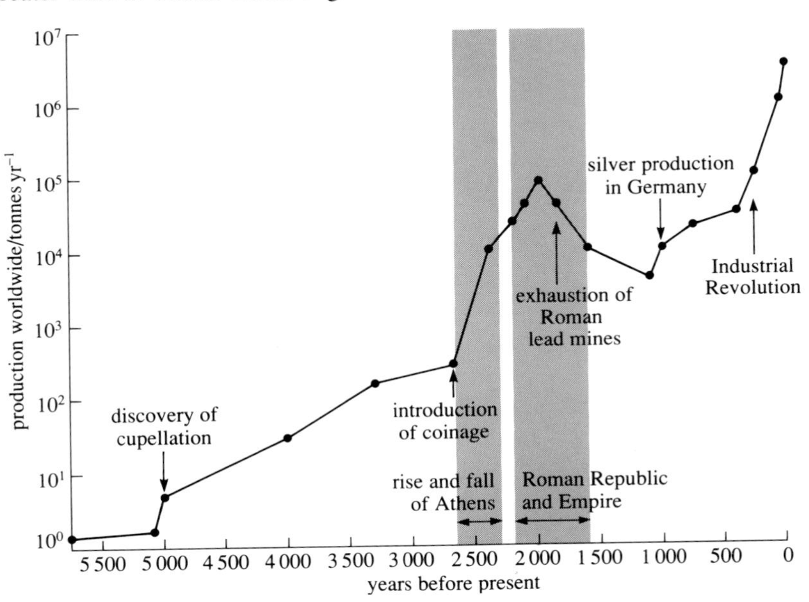

Figure 2.1 Estimated world production of metallic lead during the past 5 500 years. The scale on the horizontal axis tells you the number of years prior to the present date; the scale on the vertical axis tells you the lead production in that year. Note that the scale on the vertical axis is of the type known as logarithmic; that is, it is labelled in powers of ten, and successive powers of ten occur at equally spaced intervals along the axis. Thus the interval at 10^1 corresponds to 10 tonnes per year, the second at 10^2 to 100 tonnes per year, the third at 10^3 to 1 000 tonnes per year and so on. Cupellation was a technique for extracting silver from the silver impurities in metallic lead. Its discovery made lead smelting more attractive.

Activity 2.1

Check that you can extract data from Figure 2.1. Read the caption and then work out (a) at what date lead production peaked during the Roman period, and (b) what production was in that year. Give your answer to (b) as a number, not as a power of ten.

Today, metallic lead is used primarily in electrical storage batteries such as the lead car battery, in solders and as sheet, piping and cable sheathing in the construction industry. In the heyday of leaded petrol, around 1974, some 20–25% of lead produced was converted into petrol additives. This trade has now shrunk dramatically, but metallic lead can be turned into other important compounds, and some of these are now described.

2.1 Reactions of lead and some lead compounds

Lead oxide, PbO, is an orange solid which is made commercially by blowing air through molten lead. It has long been added to some glass-forming materials because it produces a brilliant tone in pottery glazes and in bulk glass such as the famous 'lead crystal' type. It dissolves in dilute nitric or acetic acids by reacting with hydrogen ions (Figure 1.4) to give the colourless aqueous ion, $Pb^{2+}(aq)$, which is the most common form of lead in aqueous solution.

▷ Write the equation for the reaction of lead oxide, PbO(s), with aqueous hydrogen ions, $H^+(aq)$, which gives $Pb^{2+}(aq)$.

▶ See Figure 1.4; the reaction is

$$PbO(s) + 2H^+(aq) \longrightarrow Pb^{2+}(aq) + H_2O(l) \qquad (2.3)$$

The word *plumbing* marks the fact that metallic lead (Latin: *plumbum*) was formerly used to make pipes and cisterns which carried or stored drinking water. This is safest with hard water which contains aqueous ions such as sulphate, $SO_4^{2-}(aq)$, and carbonate, $CO_3^{2-}(aq)$ (Table 1.2), because the metal then becomes covered with thin films of solids such as lead sulphate, $PbSO_4$, lead carbonate, $PbCO_3$, and lead hydroxide, $Pb(OH)_2$. These solids have a very low solubility in water and protect the metal from corrosion. However, if the water is soft and slightly acidic, there is less protection, and lead can dissolve to reach unacceptable concentrations through a reaction involving atmospheric oxygen:

$$2Pb(s) + O_2(g) + 4H^+(aq) \longrightarrow 2Pb^{2+}(aq) + 2H_2O(l) \qquad (2.4)$$

The tendency of lead to dissolve is especially marked if it is exposed to fruit juices or alcoholic drinks which, besides being mildly acidic, contain other substances which encourage the dissolving of lead. Fruit juices, wines and spirits also have a tendency to slowly dissolve some lead oxide out of pottery glazes and leaded glass.

When lead sheet is exposed, in a confined space, to vinegar and air which has been enriched in carbon dioxide, the lead is corroded, and becomes encrusted with a brilliant white deposit known as white lead. Its empirical formula shows it to be a combination of lead carbonate and lead hydroxide in the ratio 2 : 1, so it is written $[2PbCO_3.Pb(OH)_2]$. Until 1940, when it began to be replaced by titanium dioxide, white lead was the standard pigment used in white paint. Other lead pigments used in paints include the rust-red oxide, Pb_3O_4, and yellow lead chromate, $PbCrO_4$.

2.2 Lead poisoning

Substances commonly termed poisonous are those which have harmful or fatal effects when inadvertently swallowed or inhaled in quite small amounts. Lead compounds fall into this category, and this has been known for a long time. Nikander gave a very clear account of the effects of poisoning by white lead in the 2nd century BC. He noted pallor, colic or bellyache, a constriction of the palate and gums, irritability, dangling paralytic hands, torpor, distorted vision, hiccups, a dry cough, nausea and death. Not surprisingly, therefore, the health of some society women was not improved by the long-standing habit of using white lead to acquire a pale complexion (Figure 2.2).

Other historical cases of lead poisoning arose from the habit of storing or even heating up grape or other fruit juices in leaden pots. This practice was widespread in Roman times. In the presence of oxygen, a little lead dissolved in the slightly acid juices and this inhibited enzyme activity, so acting as a preservative and preventing fermentation. Nor did lead impair the flavour, because solutions containing negative ions such as acetate, $CH_3CO_2^-(aq)$ (which are present in wine) together with lead ions, $Pb^{2+}(aq)$, are sweet.

In the 18th century there were several epidemics in the West Country of what became known as Devonshire colic. The symptoms included bellyache and paralysis. Eventually it was noticed that the disease was confined to villages where cider was prepared in presses and vats lined with lead. In 1767, the Queen's physician, George Baker, extracted 3 milligrams of lead from a pint of Devonshire cider; since some farm labourers reportedly drank a gallon a day, their lead intake from this source alone would be nearly 200 times that of today's heavy drinker! Even nowadays, some moonshine whisky drinkers in the USA fall victim to lead poisoning because they connect up their distillation tubing with lead solder, and use scrap car radiators containing lead as condensers. Imported earthenware with glazes containing illegally high concentrations of lead has also been responsible for lead poisoning in the USA. One case that led to the death of a child was traced to an earthenware jug of apple juice which was constantly topped up.

Figure 2.2 Mary Gunning, Lady Coventry (1733–1760). The most fashionable beauty of her day, she insisted, despite the known dangers, on using white lead as a cosmetic, and this contributed to her early death. In 1766, the author Horace Walpole wrote, 'that pretty young woman, Lady Fortrose, Lady Harrington's eldest daughter, is at the point of death, killed, like Lady Coventry and others, by white lead, of which nothing could break her'.

Another currently endangered group is that of children aged 2–5 who live in old, dilapidated housing and have the habit of eating non-food material such as chips of paint and putty. This behaviour is called **pica**, from the Latin word for magpie. If the old, peeling paint contains white lead or other lead pigments, these children are obviously at risk.

The symptoms of lead poisoning include colic, constipation, pallor, anaemia, wrist and foot drop, kidney damage and loss of appetite. There may also be a blue line around the gums, a long recognized mark of the careless potter using lead glazes. Finally, and perhaps more important than these physiological symptoms, there is **encephalopathy.** This is a mental or nervous disorder made manifest by headaches, irritability, insomnia, apprehension, confusion, nightmares and fits. Recovery from encephalopathy is often incomplete, and residual brain damage is common.

2.3 The measurement of lead

If we are to assess any potential threat from environmental lead, we must be able to make accurate measurements of lead levels in people and in the environment. Modern analytical chemistry cannot only do this; it can also do it quickly, so that each day many samples can be processed by a single instrument. The most common technique used for measurements of this sort is **atomic absorption spectroscopy (AAS)**, and in this section you will see how it might be used to determine the amount of lead in a sample of apple juice. The basic principles of AAS are those of the quantum theory of the atom, and Box 2.1 summarizes what you need to know about this subject.

Box 2.1 Quantum energy states and atomic spectroscopy

2.1

The free, uncombined atom or ion of a chemical element consists of a positively charged nucleus surrounded by negatively charged particles called electrons. The distribution of the electrons about the nucleus is usually the one with the lowest possible energy, and when this is so, the atom or ion is said to be in its electronic ground state.

Apart from the ground state, there are also states of higher energy in which the electrons are distributed about the nucleus in other ways. There are, however, only a limited number of these higher energy or **excited states**, and therefore only a limited set of higher energy levels which are open to the atom or ion: most energy values above the ground state energy cannot be adopted by the atom or ion. Consequently, if the atom or ion is in its ground state, and it then takes up energy from some source, it must be just the

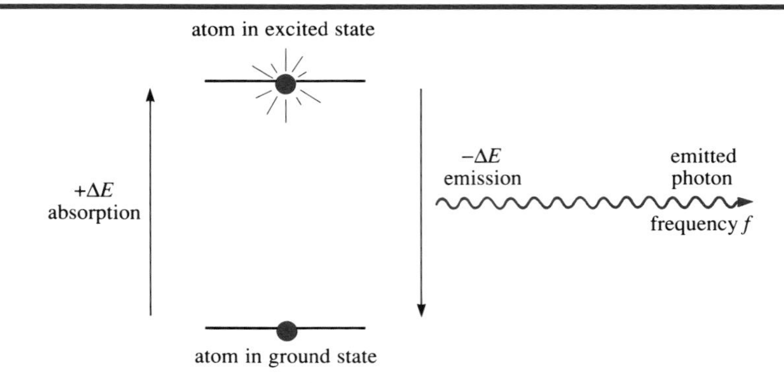

Figure 2.3 The absorption of energy and subsequent emission of radiation by an atom.

exact amount of energy needed to transfer it to one of its limited number of excited states. This process, shown on the left of Figure 2.3, is known as *absorption*.

Now that the atom is in an excited state, the reverse process of *emission* can take place: the atom can fall back to its ground state; but the energy which it loses in the process cannot be destroyed: it is transformed into a photon of radiation which the atom emits. The energy difference, ΔE, between the two states is related to the frequency, f, of the emitted radiation by the Einstein relation:

$$\Delta E = hf \qquad (2.5)$$

Where h is Planck's constant and has the value 6.626×10^{-34} J s.

Such events occur when you throw salt into a gas flame. The hot flame produces free sodium atoms, many of which are in an excited electronic state whose energy exceeds that of the ground state by that of a photon of orange light. As these excited sodium atoms quickly fall back to the ground state, such photons are emitted, colouring the flame orange. In this and other cases, the emitted radiation is characteristic of the chemical element in question. It can therefore be used to tell us if a particular chemical element is present, but atomic absorption spectroscopy goes further, and tells us how much. ■

Question 2.1 The excited state of the sodium atom which gives rise to the orange light that we have just described, lies 3.371×10^{-19} J above the ground state. What is the frequency of the orange light?

Figure 2.4 demonstrates the basic principles of lead measurement with one modern type of atomic absorption spectrometer. The centre piece is the electrothermal atomizer. It consists of a graphite platform seated inside a graphite tube through which a large electric current can be passed. In the top of the tube is a small hole through which a microsyringe can deliver a small sample of our lead-bearing apple juice on to the platform. The electric current is then used to heat the platform quickly in three short stages. In the first stage, at about 150 °C, the sample is dried out by evaporation of the water; in the second, at perhaps 600 °C, it loses any volatile organic material and is charred to an ash; in the third, the temperature is raised to over 2000 °C, when the lead and other elements are vaporized and converted to free atoms, i.e. atomized.

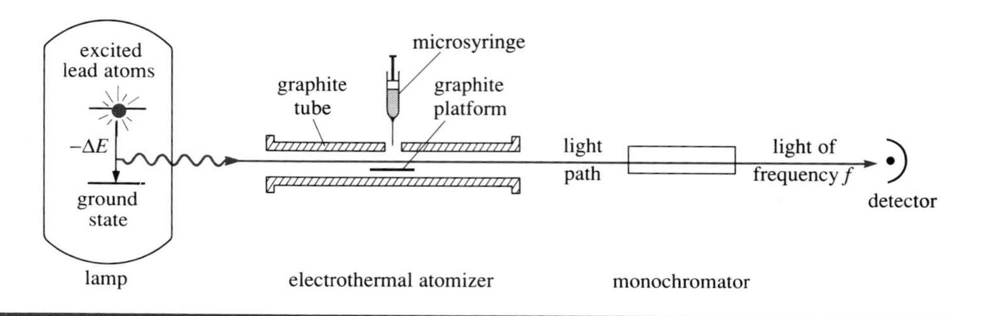

Figure 2.4 An atomic absorption spectrometer using electrothermal atomization.

The lead-bearing vapour lies in the path of radiation arising from a source of lead atoms in excited electronic states. These are produced in a special lamp, by bombarding a target made of metallic lead with fast-moving noble gas ions such as neon, Ne^+. Let us consider lead atoms in a particular excited energy state; they will soon revert to the ground state, and as they do so they will emit light of frequency f. If there is no vapour above the graphite platform, the light will pass through the graphite tube without losing any intensity.

▷ What happens if there is a vapour containing lead atoms above the platform?

▶ The radiation of frequency f will have just the energy needed to lift the lead atoms to the excited state that produced this radiation in the lamp. Some of it will be absorbed by the lead atoms, and so the intensity of the radiation will be reduced.

The amount of radiation absorbed depends on the concentration of lead atoms in the vapour, and therefore on the amount of lead in the original sample. It can therefore be used to measure that amount of lead.

The radiation leaving the tube passes through a device called a monochromator. This simply blocks out all radiation except that with the frequency, f, emitted by our chosen excited state of the lead atom. A suitable excited state emits ultraviolet radiation with a frequency of 1.058×10^{15} Hz. It passes to a detector connected to a recording instrument which shows, either graphically or by calculation, the extent to which the radiation has been absorbed by the lead-bearing vapour.

A very convenient measure of the amount of radiation that is absorbed is a quantity called the **absorbance**, because this is proportional to the concentration of the absorbing atoms in the sample. A concentration tells us the amount of a substance in a particular volume of space. We want the concentration of lead in our sample of apple juice, and this is conveniently expressed by an amount of lead (the mass in micrograms) in a specified volume (one millilitre) of the juice. Thus in this case, the units of concentration are micrograms per millilitre ($\mu g\, ml^{-1}$). If you have difficulty with the concept of concentration or its units, there is further help in Section 2.7.

Figure 2.5 shows how the fact that absorbance is proportional to concentration can be used to determine the concentration of lead in the apple juice.

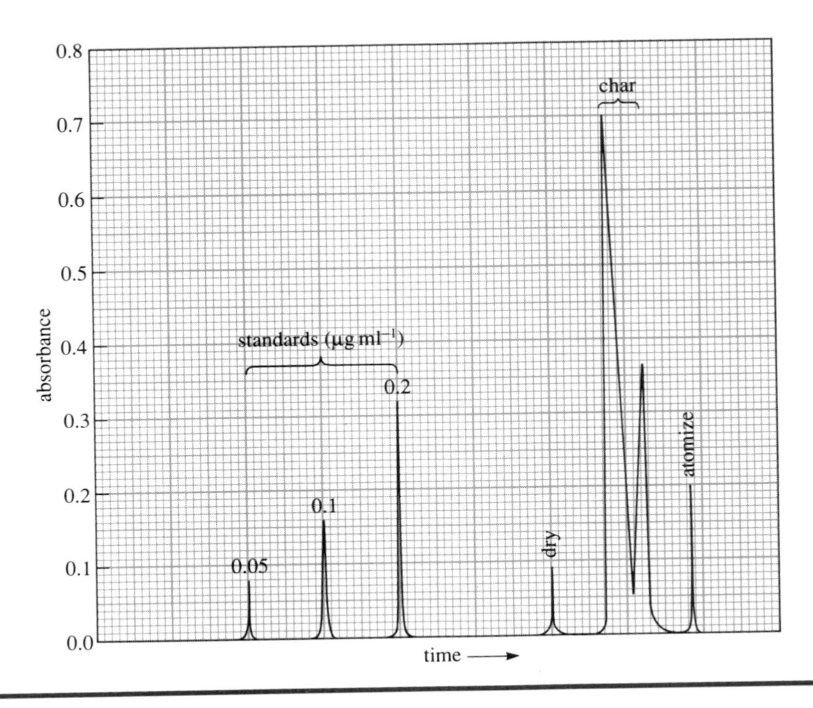

Figure 2.5 Typical output from an atomic absorption spectrometer, fitted with an electrothermal atomizer, which is being used to determine the concentration of lead in apple juice. Absorbance is measured on the vertical axis; notice that absorbances have no units; they are merely numbers.

Starting on the left of Figure 2.5, the peak heights of the first three signals record the absorbances of three carefully prepared standards. These were solutions of lead nitrate in water containing 0.05, 0.1 and 0.2 micrograms of lead per millilitre of solution ($\mu g\,ml^{-1}$). Samples of these solutions with a volume of 2 microlitres ($2\,\mu l$) were dispensed in turn on to the platform of the electrothermal atomizer.

▷ How do the three signals bear out our claim that the absorbance of a sample is proportional to its concentration?

▶ The absorbances (peak heights) are in the same ratios as the concentrations. Thus the third peak is twice as high as the second, and the second is twice as high as the first.

The remaining peaks record the absorbance of a $2\,\mu l$ sample of the apple juice. In this case, the sample also contains organic materials, and these give rise to vapours which absorb radiation of the chosen frequency during both the drying and charring stages. This is the origin of the peak marked 'dry' and the two peaks marked 'char'. Only during the last, atomization, stage do the lead atoms get into the light beam, so the final peak on the right is the one that yields the lead concentration.

▷ Use this peak and that for the standard of concentration $0.2\,\mu g\,ml^{-1}$ to calculate the concentration of lead in the apple juice.

▶ $0.125\,\mu g\,ml^{-1}$.

Because absorbance is proportional to concentration:

$$\frac{\text{concentration of lead in juice}}{\text{concentration of lead in standard}} = \frac{\text{absorbance for juice}}{\text{absorbance for standard}}$$

Now the absorbance of the apple juice is the height of the peak marked 'atomize' when read off on the vertical axis; this is 0.20. Likewise, for the standard of concentration $0.2\,\mu g\,ml^{-1}$, this height or absorbance is 0.32.

$$\text{Thus} \quad \frac{\text{concentration of lead in juice}}{0.2\,\mu g\,ml^{-1}} = \frac{0.20}{0.32}$$

$$\text{Therefore concentration of lead in juice} = \frac{0.20}{0.32} \times 0.2\,\mu g\,ml^{-1}$$

$$= 0.125\,\mu g\,ml^{-1}$$

Thus AAS can be used to determine lead levels in commercial drinks; how about lead levels in people?

2.4 Lead in people

Let us first ask why measurements of lead levels in people might be useful. Section 2.2 described the human body's reponse to very high doses of lead. The symptoms were those of **frank lead poisoning**. By frank poisoning we mean poisoning whose symptoms are so marked that they are recognizable by an intelligent but untrained person. Those symptoms are usually taken to be encephalopathy, colic and marked anaemia. The victims of such poisoning have unusually high levels of lead in their bodies, and if those levels can be determined, then the results may help us in choosing values that can be used to signal a risk of frank lead poisoning in other cases, and to trigger action that will reduce the person's exposure to lead. As you will see, this is not quite as easy as it sounds, but nevertheless, results of this kind will be especially useful in the glass, pottery and lead-smelting industries where workers are at particular risk of what is called **chronic exposure**. This is sustained exposure to a high-lead environment over periods of time. It is to be distinguished from **acute exposure** which occurs in a short space of time.

▷ In 1988, a report in the medical journal *The Lancet* claimed that the painters Rubens, Renoir and Dufy suffered from a rheumatic disease caused by lead that they had ingested from their brightly coloured lead-based paints. If true, was this a case of acute or chronic lead-poisoning?

▶ Of chronic poisoning; the rheumatic disease was presumably brought on by exposure over many years of painting; a sudden large dose was not responsible.

One reason why the distinction between acute and chronic exposure is important is that the chronically exposed acquire a certain degree of tolerance that the general population lacks. They can carry higher lead levels in their bodies before the symptoms of frank poisoning appear.

Frank lead poisoning is, of course, very rare, but lead is a very widely used element and everybody contains some. Once one has chosen lead levels in the body which signal a risk of frank lead poisoning, then it becomes important to find how far levels in the general population, and certain groups within it, fall below these values. Studies of this sort also provide opportunities for finding out if lead affects the human body at levels below those at which there is risk of frank lead poisoning.

With this background, let us consider the average person's approximate daily lead intake. It may come from a variety of sources. On average, there is about 15 g of lead in every tonne of the Earth's crust, and all soils contain some lead which may enter the food chain through uptake by edible plants. Airborne lead, including that from car exhausts, may be inhaled or deposited on plant leaves. Tinplate usually contains small quantities of lead, so canned food may contain lead from the tin, or from lead solders if these have been used to seal the can. Water takes up lead from soils and plumbing systems. The total intake can be studied by AAS and other analytical techniques, and a typical result is shown in Figure 2.6 and its caption. Measurements of lead in people are usually made on the bones, blood, urine, hair or teeth. As Figure 2.6 shows, most absorbed lead ends up in the skeleton, so bone lead is the best measure of a person's lead burden, the total amount of lead that he or she contains. It is also the best index of lifetime exposure to lead. Until quite recently, the determination of bone lead in living people was not possible, but it can now be done using a technique called X-ray fluorescence. However, this is relatively costly and time-consuming, so the method is not suitable for large-scale surveys, and here, we do not discuss it further.

The measurement of lead in teeth comes closest to being a substitute for measurements on bone. It can be done by dissolving weighed amounts of tooth in nitric acid, diluting to an accurately known final volume, and then measuring the concentration of lead in the solution by AAS, as described in Section 2.3. Measurements made on the shed milk teeth of 6–7-year-olds have been especially useful in assessing how much lead has been taken up by children. However, teeth are only available from subjects at special times, so they are not suitable material for large general surveys.

This brings us to lead in blood. As Figure 2.6 implies, blood is where most lead is parked on its way to the bones. Because the lead that reaches the blood typically spends weeks or months there, rather than years, blood lead concentrations are a measure of relatively recent exposure. The measurements can be carried out quickly, and in large numbers by AAS. Indeed, apart from the fact that special chemicals must be added to blood prior to the measurement, the method is very similar to the one

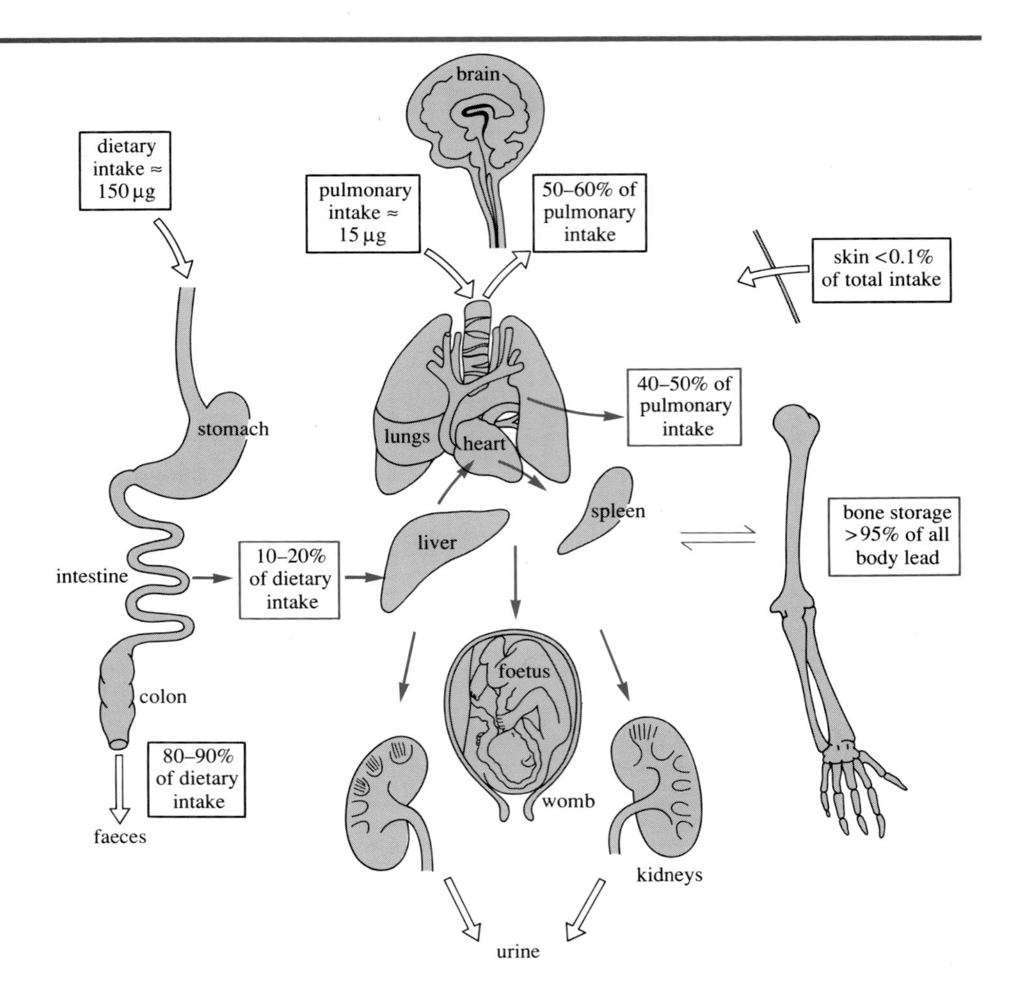

Figure 2.6 Uptake and fate of lead in a woman: about 150 μg are ingested (eaten and drunk) each day, and about 15 μg are inhaled. Most lead is therefore dietary and ingested, but only 10–20% of this is absorbed, the other 80–90% mostly passing out in the faeces. The inhaled intake is smaller, but the percentage uptake of 40–50% is larger, the balance being exhaled. The lead absorbed from either source is transported first by blood (red arrows) to soft tissues where it typically spends several weeks before moving on, and subsequently to the reservoir of the skeleton where the residence time is much longer (years rather than months). More than 95% of the total body-lead is found in bone as lead phosphate.

used for apple juice in Section 2.3. For these reasons, most information about lead levels in people has been obtained in the form of blood lead concentration, and this is also the quantity used in framing legislation. This explains why blood lead concentrations are strongly emphasized in the rest of this book.

2.4.1 Lead levels in human blood

Lead concentrations in blood are usually expressed as the number of micrograms of lead in 100 millilitres of blood. As 100 millilitres is one decilitre (1 dl), the concentration units are $\mu g\,dl^{-1}$. In the late 1970s, EEC* Directive 77/312/EEC required member states to undertake surveys of blood lead in urban dwellers, and in groups particularly exposed to lead pollution in their jobs. The Directive specified blood lead levels which, if exceeded, should trigger action to reduce exposure: no more than 50% of any group should have blood lead concentrations above $20\,\mu g\,dl^{-1}$, no more than 10% above $30\,\mu g\,dl^{-1}$, and no more than 2% above $35\,\mu g\,dl^{-1}$. What governs this sort of choice is discussed later in the section.

The UK surveys took place in 1979 and 1981. Most of the groups studied had blood lead levels which did not breach the Directive. In inner cities in the UK, the average concentration was $13\,\mu g\,dl^{-1}$, and only 9.3% of the people investigated had blood lead levels in excess of $20\,\mu g\,dl^{-1}$. The situation was worst in inner Manchester where the average was $17\,\mu g\,dl^{-1}$. In Figure 2.7, the blood lead concentrations of the sample of inner Manchester men are shown in the form of a histogram. On the horizontal axis,

* European Economic Community; now called simply the European Community (EC).

these concentrations are broken down into intervals of 5 μg dl^{-1}. From each interval springs a vertical bar whose height, when measured against the scale on the vertical axis to the left, gives the percentage of the sample whose blood lead falls within the interval in question. Now try Activity 2.2.

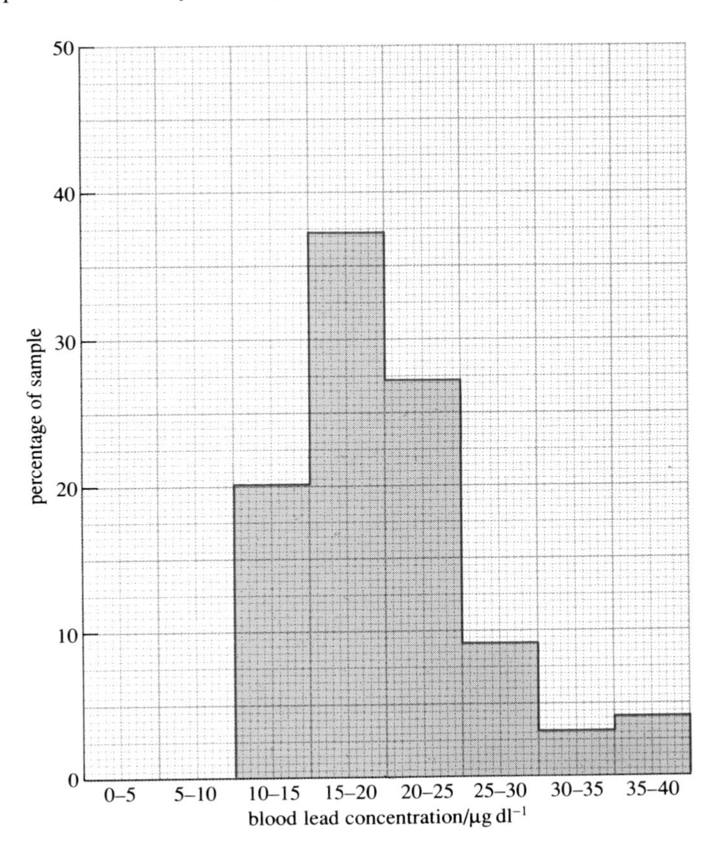

Figure 2.7 The distribution of the blood lead concentrations in a sample of inner Manchester men in 1979.

Activity 2.2

Roughly what percentage of the inner Manchester sample of Figure 2.7 had blood lead concentrations (a) above 35 μg dl^{-1} and (b) above 20 μg dl^{-1}? Do either of your calculated percentages breach the EEC Directive cited above?

The 35 μg dl^{-1} percentage limit in the EEC Directive was also breached by adults in Islington, and by the children of lead workers at Gravesend in Kent. As a result of such surveys, the 35 μg dl^{-1} percentage limit was then effectively made redundant in the UK through a DHSS* recommendation of action on behalf of any person with a blood lead level in excess of 25 μg dl^{-1}. This recommendation, made in 1982, advised that 'where a person—particularly a child—is confirmed as having a blood lead level over 25 μg dl^{-1}, his or her environment should be investigated for sources of lead, and steps taken to reduce exposure'.

It is interesting to compare the figures which have already been discussed in this section, with the lowest values at which the symptoms of frank lead poisoning, and less serious effects of lead exposure, have been observed. A few results from reviews of this subject are shown in Table 2.1, and some of the data for children alone are portrayed in Figure 2.8.

* Department of Health and Social Security.

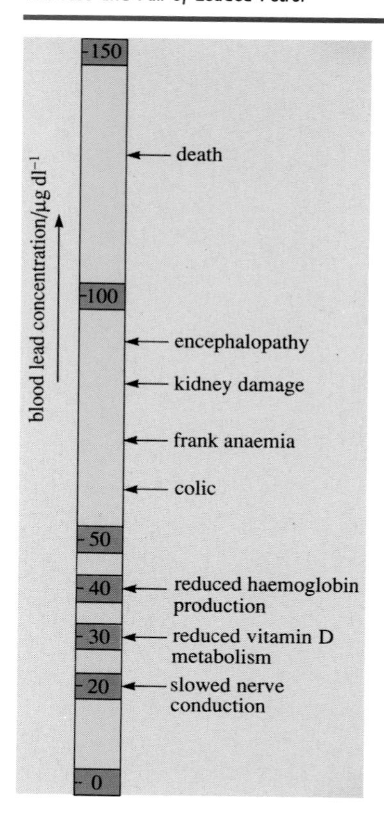

Figure 2.8 Lowest blood lead levels at which certain harmful effects of lead have been adequately demonstrated in children.

Table 2.1 Lowest observed effect levels for some lead-induced health effects. From *Air Quality Criteria for Lead*, Volume 1, Environmental Protection Agency EPA-600 (1986), and *Preventing Lead Poisoning in Young Children*, US Department of Health and Human Services (1991).

Effect	Lowest observed effect level (blood lead concentrations/$\mu g\,dl^{-1}$)	
	Adults	Children
encephalopathy	100–120	80–100
frank anaemia	80	70
colic		60
sub-encephalopathic symptoms	40–60	
reduced haemoglobin production	50	40
slowed nerve conduction	40	20

It must be emphasized that Table 2.1 and Figure 2.8 show the *lowest* blood lead concentrations at which the effects in question have been observed. It is not possible to identify blood lead levels at which a particular person will begin to suffer from, say, encephalopathy because individuals vary greatly in their resistance to harm from lead. Some children with blood lead concentrations of $200\,\mu g\,dl^{-1}$ have shown no obvious clinical symptoms, whereas others have suffered irreversible brain damage, and even death, at levels of about $130\,\mu g\,dl^{-1}$. Nevertheless, Table 2.1 clearly implies that frank lead poisoning may sometimes be associated with blood lead concentrations as low as $80\,\mu g\,dl^{-1}$ in adults.

▷ What is the corresponding figure for children?

▶ 60–70 $\mu g\,dl^{-1}$; colic and frank anaemia are symptoms of frank lead poisoning and, according to Table 2.1, have been observed at this level.

Table 2.1 and Figure 2.8 also show that, well below such levels, there may be other effects such as slowed nerve conduction, lowered vitamin D metabolism and reduced haemoglobin production. These are not serious enough to be classed as symptoms of lead poisoning, but they clearly provide food for thought.

What this tells us is that lead levels that are dangerous for some people seem relatively safe for others. Consequently, a nation's choice of 'danger levels' which call for remedial action are not just a matter of scientific measurement and simple logic. They will vary from one country to another, reflecting internal opinions about what resources can be spared, and whether they should be allocated to the lead problem, rather than to many other worthy calls on the national purse. Clearly then, 'danger levels' are likely to change with time, and to be different in, say, the UK and Romania. In the UK, it is generally agreed that a blood lead concentration of $70\,\mu g\,dl^{-1}$ is uncomfortably close to levels at which frank lead poisoning may occur. At this level, a male lead worker must be suspended from work unless his length of service in the lead industry and other tests can provide indications of special tolerance. For female workers, the suspension figure is only $40\,\mu g\,dl^{-1}$. It is lower for women because lead crosses the placental barrier (see Figure 2.6), so pregnant women need special safeguards. Prior to 1980, the male worker threshold in the UK was $100\,\mu g\,dl^{-1}$; in the USA, where environmental lead has received much publicity, the figure is $50\,\mu g\,dl^{-1}$ for both sexes. Thus 'danger levels' are a matter of social choice. But notice that without analytical chemistry, that choice could not be made!

2.5 Lead levels in air

Lead in air is determined by drawing air through a filter with a pump. The pump measures the volume of air, and the suspended lead particles are deposited on the filter. They can subsequently be dissolved in acid, made up to a known volume with water, and the dissolved lead measured by AAS (Section 2.3).

The concentration of lead in air is usually expressed in micrograms per cubic metre ($\mu g\,m^{-3}$). If you have difficulty with these units of concentration, you will find help in Section 2.7. In rural areas in the UK, the average lead content of air is usually below $0.15\,\mu g\,m^{-3}$; in most cities, below $1\,\mu g\,m^{-3}$. During the summer of 1978, however, a mean value of nearly $10\,\mu g\,m^{-3}$ was found on the central reservation of the M4 motorway, and Figure 2.9 shows how this concentration changed with distance from the motorway and wind direction. The peak concentration, of over $15\,\mu g\,m^{-3}$, occurred about 15 m downwind of the central reservation. Under EEC Directive 82/844/EEC, member states were obliged to ensure that, from January 1988, the annual average lead concentration in the air at any location does not exceed $2\,\mu g\,m^{-3}$.

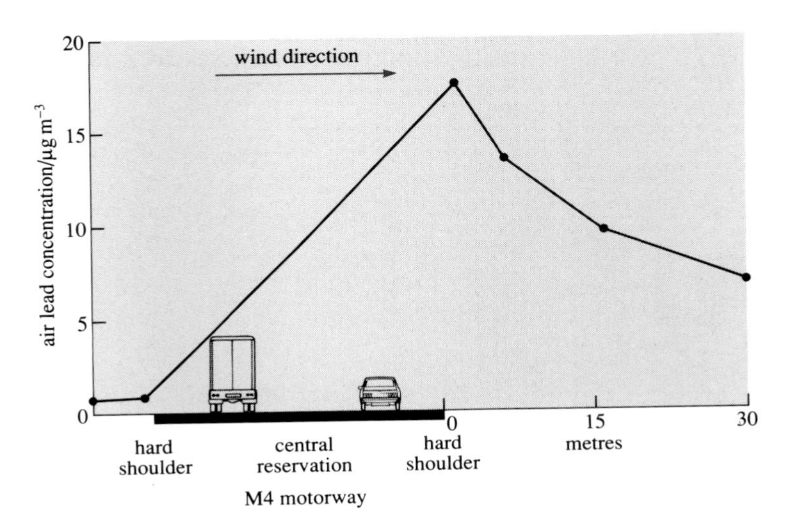

Figure 2.9 Mean lead concentrations in air on the busy M4 motorway in the summer of 1978. Note the effect of wind on the distribution.

2.6 Lead levels in tap-water

The daytime tap-water in most households in the UK has a lead concentration of less than 10 micrograms per litre ($10\,\mu g\,l^{-1}$). However, a survey in the mid-1970s noted values as high as $1\,000\,\mu g\,l^{-1}$, and in the Glasgow area, where water was soft, slightly acid, and delivered through lead pipes, over 50% of households had concentrations well above $100\,\mu g\,l^{-1}$, the mean value being $244\,\mu g\,l^{-1}$. EEC Directive 80/778/EEC prescribes an upper limit of $50\,\mu g\,l^{-1}$. The situation in Glasgow has since been improved by encouraging the replacement of lead plumbing, and by reducing the acidity of the water through the addition of lime (calcium hydroxide, $Ca(OH)_2$). This acts as a source of hydroxide ions, $OH^-(aq)$, which can neutralize $H^+(aq)$:

$$Ca(OH)_2(s) \longrightarrow Ca^{2+}(aq) + 2OH^-(aq) \qquad (2.6)$$

$$H^+(aq) + OH^-(aq) \longrightarrow H_2O(l) \qquad (2.7)$$

▷ Why should the lowering of acidity reduce the lead content of the water?

▶ Equation 2.4 and the adjacent comments show that hydrogen ions favour the dissolution of lead, especially where water is soft. The destruction of $H^+(aq)$ via the reaction shown in Equation 2.7 therefore reduces this dissolution.

As a result of such treatment, by the 1980s, over 95% of Glasgow water samples had lead concentrations of less than $100\,\mu g\,l^{-1}$.

2.7 Handling and interconverting units

In a teaching course, a scientific quantity such as a concentration is usually expressed in just one type of unit. But as Sections 2.4.1–2.6 reveal, research scientists use different units for the concentrations of lead in blood, air and water. If you found this confusing, this section is designed to help you. It does so first by building up the concept of concentration itself, and then clarifying the relationship between the different units of concentration by showing how they can be interconverted. The ability to interconvert scientific units for the same quantity is an important general skill, and the method considered here can be extended to quantities other than concentration.

The conventional measure of the amount of a substance is its mass, which can be expressed in grams (symbol g). Likewise, volumes can be expressed in litres (symbol l). From units of this kind, the units of other quantities can be built up. The concentration of lead in a liquid sample, for example, is usually expressed as the amount of lead in a specified volume of the liquid. If the volume of the sample is 2 litres, and it contains 4 grams of lead, then each litre contains 2 grams, and the concentration of lead is 2 grams per litre, written $2\,g\,l^{-1}$.

Table 2.2 The names, numerical values and symbols for the fractional parts which you will find attached to units in this book.

Name	Symbol	Value
deci	d	10^{-1}
centi	c	10^{-2}
milli	m	10^{-3}
micro	μ	10^{-6}
nano	n	10^{-9}
pico	p	10^{-12}

Sometimes, concentrations are so small that these units lead to very clumsy expressions. The concentration of lead in a sample of tap-water may be $0.000\,009\,6\,g\,l^{-1}$. By using powers of ten, the concentration can be written $9.6 \times 10^{-6}\,g\,l^{-1}$. This is an improvement, but it can be made even more concise by replacing the units by their fractional parts (one-tenth, one-thousandth, one-millionth etc.). Some of the fractional parts used in this book, along with names, numerical values and symbols are given in Table 2.2.

As an illustration, let us convert the concentration of $0.000\,009\,6\,g\,l^{-1}$ to a more manageable form. (This is a very simple example, but is used here to demonstrate the technique.) As already noted, this is $9.6 \times 10^{-6}\,g\,l^{-1}$. Now from Table 2.2,

$$1 \text{ microgram} = 10^{-6} \text{ grams}$$
$$\text{or } 1\,\mu g = 10^{-6}\,g$$

We now produce a *conversion factor which is equal to one*, by dividing both sides of this equation by $10^{-6}\,g$:

$$\frac{1\,\mu g}{10^{-6}\,g} = 1$$

Since this is equal to one, our concentration of $9.6 \times 10^{-6}\,g\,l^{-1}$ will be unchanged if we multiply it by the conversion factor. Thus:

$$\text{concentration of solution} = (9.6 \times 10^{-6}\,g\,l^{-1}) \times \frac{1\,\mu g}{10^{-6}\,g}$$
$$= 9.6\,\mu g\,l^{-1}$$

Study carefully what we have done here. We needed to eliminate the g in $9.6 \times 10^{-6}\,g\,l^{-1}$. Thus the conversion factor had to be arranged so that when $9.6 \times 10^{-6}\,g\,l^{-1}$ was multiplied by it, the g symbols cancelled out. This meant that the g symbol had to appear on the bottom of the conversion factor. If cancellation had required the g symbol to appear on the top, then because the conversion factor is equal to one, it could simply have been written the other way up:

$$\frac{10^{-6}\,g}{1\,\mu g} = 1$$

When converting quantities into different units by this method, it is essential to make the right choice between these two ways of writing the conversion factor.

The same procedure can be used in more complicated cases, as described below.

The concentration of lead in air is typically $0.000\,000\,000\,15\,\text{g}\,\text{l}^{-1}$.

▷ First of all, express this in powers of ten notation.

▶ The concentration is $1.5 \times 10^{-10}\,\text{g}\,\text{l}^{-1}$.

Now 1 cubic metre = 10^3 litres ($1\,\text{m}^3 = 10^3\,\text{l}$).

So we can write the conversion factor:

$$\frac{10^3\,\text{l}}{1\,\text{m}^3} = 1$$

The conversion factor has been written this way up because we shall be eliminating the l (litres) symbol from $1.5 \times 10^{-10}\,\text{g}\,\text{l}^{-1}$. Thus the l (litres) symbol must be on top of the conversion factor because when we multiply by it, the litres cancel out:

$$\text{litres} \times \text{litres}^{-1} = \text{litres} \div \text{litres} = 1$$

▷ Now use this conversion factor, and the grams to micrograms conversion factor to convert the concentration of lead in air to micrograms per cubic metre ($\mu\text{g}\,\text{m}^{-3}$).

▶ Both conversion factors are equal to one, so the concentration will be unchanged when it is multiplied by both of them:

$$\text{concentration} = 1.5 \times 10^{-10}\,\text{g}\,\text{l}^{-1} \times \frac{1\,\mu\text{g}}{10^{-6}\,\text{g}} \times \frac{10^3\,\text{l}}{1\,\text{m}^3}$$

Cancelling the symbols for the litre (l) and the gram (g):

$$\text{concentration} = 1.5 \times 10^{-10} \times \frac{1\,\mu\text{g}}{10^{-6}} \times \frac{10^3}{1\,\text{m}^3}$$

Tidying up the powers of ten:

$$\text{concentration} = 1.5 \times 10^{-1}\,\mu\text{g}\,\text{m}^{-3}$$
$$= 0.15\,\mu\text{g}\,\text{m}^{-3}$$

This more economical rendering reveals why $\mu\text{g}\,\text{m}^{-3}$ are the units chosen to express lead concentrations in air.

Summary of Chapters 1 and 2

1 Metallic lead is made by roasting lead sulphide, PbS, in air. Lead oxide is first formed and yields molten lead when heated with carbon.

2 Lead oxide, PbO, is used to make leaded glass, and some pottery glazes. It is made by blowing air through molten lead. The oxide dissolves in dilute nitric or acetic acid to give the aqueous lead ion, $Pb^{2+}(aq)$.

3 Lead sulphate, lead carbonate and lead hydroxide have very low solubilities in water. Thin films of such compounds protect lead from corrosion in moist air and hard water.

4 White lead, $[2PbCO_3.\ Pb(OH)_2]$, was once an important pigment in white paint. Children with pica who are exposed to old peeling paint are at risk from poisoning by white lead and other lead pigments.

5 Other groups at risk from lead poisoning include workers exposed to lead in the smelting, glazing, pottery and battery industries, and imbibers of alcoholic or fruit drinks stored in, or prepared in contact with, metallic lead or lead glazes.

6 Symptoms of lead poisoning include colic, anaemia, pallor, wrist and foot drop, and encephalopathy, a mental and nervous disorder.

7 Lead concentrations in a solution such as blood can be determined by atomic absorption spectroscopy. A sample is atomized in the path of radiation emitted by excited lead atoms in a lamp. The amount of radiation absorbed by the sample atoms is a measure of the lead concentration in the sample solution.

8 Most ingested or inhaled lead is excreted. The absorbed part passes to the skeleton via the blood and soft tissues. Over 90% of body lead is in the bones, and bone lead levels are a measure of long-term exposure. Blood lead is a measure of relatively recent exposure (a matter of months).

9 There is evidence that symptoms of frank lead poisoning have sometimes been observed at blood lead concentrations in the 60–100 $\mu g\, dl^{-1}$ range. In the UK, male and female employees who work with lead must be suspended at blood lead levels of 70 $\mu g\, dl^{-1}$ and 40 $\mu g\, dl^{-1}$ respectively. As for the general population, the DHSS recommended that steps should be taken to reduce the exposure of anyone whose blood lead level exceeds 25 $\mu g\, dl^{-1}$.

10 EC Directives specify upper limits of 50 $\mu g\, l^{-1}$ for lead concentrations in tap-water, and 2 $\mu g\, m^{-3}$ for the average annual lead concentration in air.

Question 2.2 Write balanced chemical equations for the following changes:

(a) When black lead sulphide, PbS(s), and white lead sulphate, $PbSO_4$(s), are heated together, molten lead and sulphur dioxide gas are formed.

(b) White lead carbonate, $PbCO_3$(s), dissolves in dilute nitric acid by a reaction with hydrogen ions, H^+(aq), to give Pb^{2+}(aq), carbon dioxide, CO_2(g), and water.

Question 2.3 A standard sample of blood with a lead concentration of 0.4 $\mu g\, ml^{-1}$, and a blood sample from a male lead worker, are treated in identical ways. Then 2 μl samples of each are dispensed in turn onto the platform of an electrothermal atomizer in an atomic absorption spectrometer. Figure 2.10 shows the output.

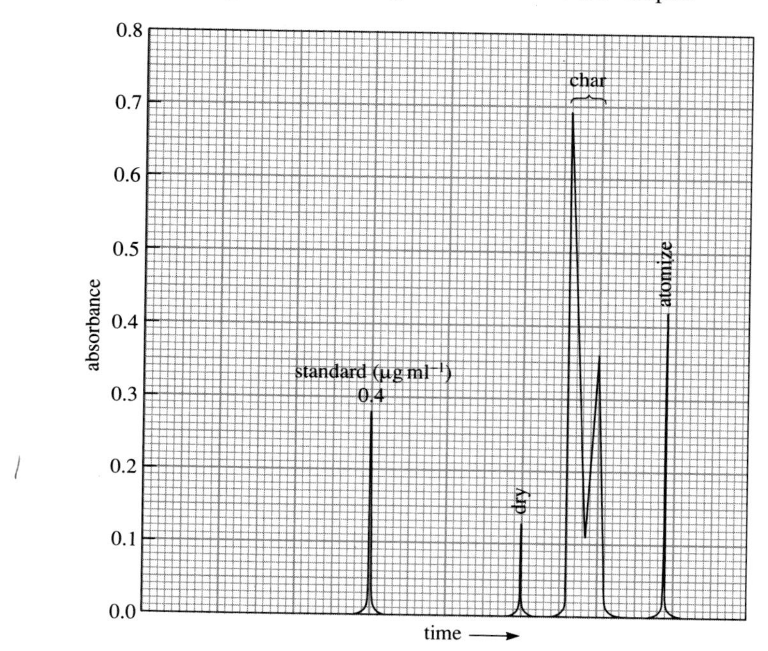

Figure 2.10 Output from an atomic absorption spectrometer. (For use with Question 2.3)

(a) Calculate the blood lead concentration of the workman.

(b) Does the result necessitate his suspension from work, (i) in the UK; (ii) in the USA?

Question 2.4 As noted in the text, the 18th century pint of Devonshire cider occasionally contained as much as 3 milligrams of lead.

(a) Given that one pint is 0.576 litres, use conversion factors to express this concentration in $\mu g\ l^{-1}$.

(b) By how much does your concentration exceed the upper limit of $50\ \mu g\ l^{-1}$ for the lead concentration in drinking water given in EEC Directive 80/778/EEC?

Activity 2.3

On 27 January 1991, the *Observer* contained an article on the British Army's new Challenger 2 tank. The calibre of the main armament was revealed in the final paragraph when the writer described how 'a laser beam tells you the range and one squeeze fires the 1 200 mm gun'. This considerably understates the drama of the event: the estimated mass of just one 1 200 mm shell would be about 30 tonnes which is approximately half that of the tank. Firing would undoubtedly be suicidal, because the recoil would shear off the turret and rip the tank apart. In actual fact, the gun calibre is 120 mm.

Numerical errors of this sort are surprisingly frequent in newspapers. When they occur, knowledge of the subject is usually all that can prevent the reader from being misled. Read Extract 2.1, and then answer the following questions.

(a) Can you spot the errors in this extract, and suggest what corrections should be made?

(b) If the extract were correct, would Athenians be perfectly healthy, at serious risk from encephalopathy, or dead?

Extract 2.1 From *New Scientist*, 31 October 1985.

Cleaning Athenian blood

THE FIRST firm evidence of a link between lead additives in petrol and lead in human blood has been discovered in Athens.

Greece has recently reduced the amount of lead allowed in petrol from 0.4 milligrams per litre to 0.15 milligrams. This produced a two-thirds reduction in the amount of lead in the air in Athens. At the same time, there has been a sharp drop in the amount of lead in the blood of Athenians. It fell in men from 18 milligrams per decilitre in 1982 to 14 milligrams in 1985, and in women from 14 milligrams to 10.

Unfortunately, the group that responded the least to the cleaner air was the group most susceptible to lead poisoning—children. The researchers are unsure whether this is because they metabolize lead differently, or because children are exposed to other sources of lead to a greater extent than their parents.

3 Lead and petrol

From Chapter 2, you know that lead is toxic, so one needs good reasons for adding lead compounds to petrol, and dispersing them in the environment. To understand what those reasons are, you need to know something of the motor car, and the engine that drives it.

3.1 The motor car and the petrol engine

The description of a petrol-driven motor car which follows is an example of scientific modelling. Modelling is useful whenever scientists single out for attention some narrow aspect of a more complicated system. To help their thinking, they construct a model. The model represents the system freed of the additional complications that the scientist has chosen to ignore. Here, we introduce a model of the motor car which consists of a chemical reaction vessel and a petrol tank. The reaction vessel is a cylinder fitted with a piston and a sparking plug (Figure 3.1).

▷ Why is such a description clearly a *model* of a motor car?

▶ Descriptions become models by virtue of the things that they ignore or omit. Here, for example, there is no mention of devices such as the crankshaft, gears, axles and wheels which transform the chemical energy made available during the ignition stroke into the kinetic energy of forward motion of the car.

The chemical reaction which supplies the energy during the ignition stroke is a **combustion reaction**: the chemical compounds in the petrol burn in the oxygen contained in the air sucked in from outside the car. For example, a typical constituent of petrol is a liquid compound called an alkane with the molecular formula C_8H_{18}. If enough oxygen is present, its combustion reaction in the cylinder is

$$2C_8H_{18}(l) + 25O_2(g) \longrightarrow 16CO_2(g) + 18H_2O(g) \quad (3.1)$$

Figure 3.1 The conventional petrol engine runs through a four-stroke cycle. First comes the down-stroke of induction (a), when the valve at the top left of the cylinder opens and the downward motion of the piston sucks in the chemical reactants, a mixture of petrol from the tank, and air from the external atmosphere. Next follows the up-stroke of compression (b), when both valves are closed and the reactants are squeezed into a smaller space. Just before the moment of maximum compression, a spark from the plug initiates the chemical reaction. This leads to ignition, the third phase of the cycle (c), which is a downward stroke of the piston propelled by the large expansion in volume associated with the explosive chemical reaction. Finally there is the upward exhaust stroke (d), when the valve at the top right of the cylinder opens and the products of the reaction are pushed out into the external atmosphere. The whole cycle can then begin again.

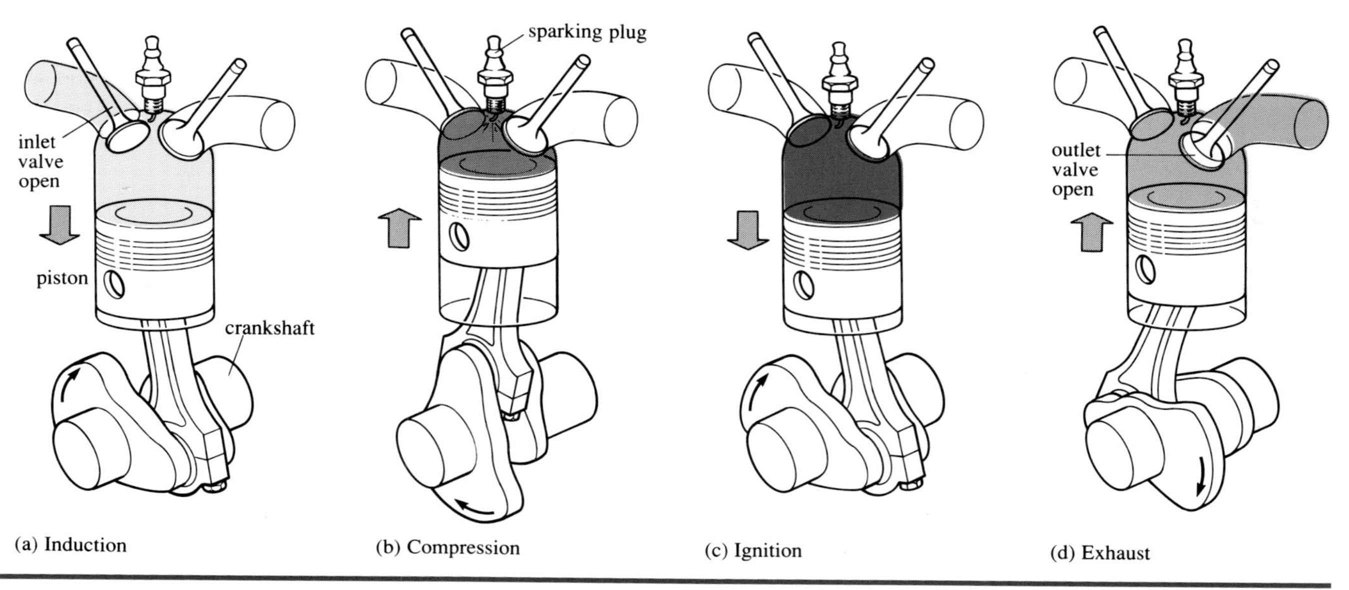

The important point about this kind of reaction is that the energy of the products is found to be much less than that of the reactants *when the comparison is made at the same temperature*. Thus, if we just think about the materials in Equation 3.1, then if the reaction were to occur without change of temperature, there would be a large decrease in energy. But the law of energy conservation tells us that energy cannot be destroyed; it must reappear in some other form.

▷ Can you guess the form in which the 'missing' energy appears?

▶ It manifests itself by an increase in temperature of the products. There is an energy decrease if the reactants and products have the same temperature, but the temperature of the products exceeds that of the reactants by just the amount needed to make the energy change zero. No energy has then been destroyed.

It is this rise in temperature that is primarily responsible for the expansion of the gases in the cylinder that pushes the piston downwards during the ignition stroke. Because the reaction occurs between gases or liquid droplets, and because the temperature rise is so great, the combustion reaction is marked by a flame. Ideally, the reactants should be consumed in a smooth movement of the flame front from the sparking plug to the furthest corners of the cylinder, and this should lead, in turn, to an agreeably smooth retreat of the piston. But sometimes this desirable movement is disrupted. The vapour in those corners of the cylinder most remote from the sparking plug is known as the *end-gas*. As the flame front advances, the end-gas ahead of it undergoes relatively prolonged heating and compression, and this can initiate chemical reactions which culminate in disorderly explosions in the end-gas before the flame front arrives (Figure 3.2). This phenomenon, and sounds such as the familiar 'pinking' noise which are associated with it, are known as **knocking**. Persistent knocking can damage the pistons, and ultimately cause engine failure.

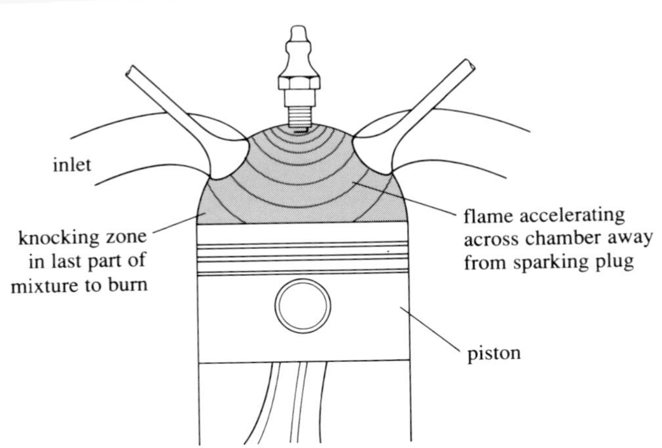

Figure 3.2 The phenomenon of engine knock occurs in the end-gas which is most remote from the sparking plug.

3.1.1 The compression ratio

One important characteristic of petrol engines is the **compression ratio**, the ratio of the volume in the cylinder at the end of the induction stroke (Figure 3.1a) just prior to compression, to the volume in the cylinder at the end of the compression stroke (Figure 3.1b). Engines with high compression ratios are desirable because they develop more power and consume less fuel. In the early 1920s, a typical compression ratio was about 4 : 1. Attempts to increase it tended to cause unacceptable engine knock because of the increased pressure on the end-gas. But oil companies have since found ways of changing the composition of petrol so that it becomes much more knock-resistant. Consequently, modern car engines can be designed with compression ratios of about 9 : 1 or 10 : 1. The use of lead additives is only one of the sources of this improvement. More important are chemical processes which destroy some substances and create new ones in the mixture of hydrocarbons that we call petrol. To understand why they are effective, you must be familiar with the structural formulae of different hydrocarbons. Box 3.1 will refresh your memory of this subject. If you do not need this reminder, pass on to Questions 3.1 and 3.2 which will check whether you are right in making that assumption!

3.1

Box 3.1 The structural formulae of hydrocarbons

Hydrocarbons consist of molecules containing just carbon and hydrogen atoms. For example, the simplest hydrocarbon is methane, the main constituent of natural gas. This consists of molecules, each with the molecular formula, CH_4.

Both crude oil and petrols contain many different hydrocarbons, each with its own unique characteristic molecule, but in every case, the numbers of carbon and hydrogen atoms within the molecule, and the way in which they are linked together, conform to two simple rules. Central to these rules is the concept of a unit **chemical bond** which holds two atoms together. This unit chemical bond is represented by a single line, and at either end are written the symbols of the elements whose atoms the bond holds together. Thus C—H represents a unit or single chemical bond which links a carbon atom and a hydrogen atom. The rules are as follows:

1 Each hydrogen atom forms just one bond which must link it to a carbon atom.

2 Each carbon atom forms four bonds, any one of which may be linked to either a hydrogen atom, or another carbon atom.

```
  H
  |
H-C-H
  |
  H

  I
```

Using these rules we can represent the molecules of hydrocarbons on paper. Thus, structure I represents methane: the four hydrogen atoms are all linked to a carbon atom by unit bonds (rule 1), and each of the four bonds formed by the single carbon atom link it, in this case, to hydrogen atoms (rule 2). Structure II shows the simplest hydrocarbon containing a link between two carbon atoms. This is ethane, C_2H_6, a minor constituent of natural gas.

```
  H H
  | |
H-C-C-H
  | |
  H H

  II
```

▷ How is rule 2 satisfied in this case?

▶ Each carbon atom forms three unit bonds to three hydrogen atoms, and a fourth to the other carbon atom.

```
  H H H H H H
  | | | | | |
H-C-C-C-C-C-C-H
  | | | | | |
  H H H H H H

      III
```

Representations of this kind are called **structural formulae** because they tell us not just the number of atoms which are present in the molecule, but also the way in which those atoms are linked to each other. We shall now see how structural formulae group hydrocarbons into classes, or families, which will later turn out to differ markedly in their tendency to cause engine knock.

Normal or straight-chain alkanes

Alkanes are hydrocarbons in which the only type of bond between two carbon atoms is the single C—C bond seen in structure II. In the normal or straight-chain alkanes, all the carbon atoms, and the single bonds between them, can be written down as a single chain. One then links hydrogen atoms to the chain of carbon atoms until rule 2 is satisfied. Structure III shows one example, which is hexane, C_6H_{14}. The structural formula consists of a CH_3 group at the left-hand end, followed by a chain of four CH_2 groups, and it then ends in another CH_3 group at the right-hand end. It therefore saves time (and space) to use the **abbreviated structural formula**, which is written

$CH_3{-}CH_2{-}CH_2{-}CH_2{-}CH_2{-}CH_3$

and is sometimes abbreviated further to $CH_3{-}(CH_2)_4{-}CH_3$, or even $CH_3(CH_2)_4CH_3$. Table 3.1 shows the abbreviated structural formulae of hexane, and two other straight-chain alkanes. All three compounds are liquids at room temperature, and their boiling temperatures are included.

▷ How do the boiling temperatures change as the size and mass of the molecule increase?

▶ As the number of atoms in the hydrocarbon molecule, and therefore its size and mass, increase, the boiling temperature increases too. This is generally the case in a family of hydrocarbons.

The molecular formula of straight-chain alkanes takes the form C_nH_{2n+2}, where n is the number of carbon atoms in the molecule. For example, octane is C_8H_{18}, the case where $n = 8$, and $(2n + 2)$ is $(2 \times 8) + 2 = 18$. You can check that hexane and heptane correspond to the cases where n is 6 and 7 respectively.

Table 3.1 The abbreviated structural formulae and boiling temperatures of hexane, heptane and octane.

Name	Abbreviated structural formula	Boiling temperature/°C
hexane	$CH_3{-}CH_2{-}CH_2{-}CH_2{-}CH_2{-}CH_3$	69
heptane	$CH_3{-}CH_2{-}CH_2{-}CH_2{-}CH_2{-}CH_2{-}CH_3$	98
octane	$CH_3{-}CH_2{-}CH_2{-}CH_2{-}CH_2{-}CH_2{-}CH_2{-}CH_3$	126

Branched-chain alkanes

Hexane (structure III) is not the only structural formula with the molecular formula C_6H_{14}. In structure III, each carbon atom is linked to either one or two other carbon atoms, but there is nothing in rule 2 which prevents this number being three or even four. However, if a structural formula is to contain carbon atoms of this kind, the carbon chain must be branched. Structures IV and V show modifications of structure III which contain carbon atoms linked to three and four other carbon atoms respectively. In both cases, rule 2 is obeyed. There should therefore be at least two other compounds, apart from hexane, with the chemical formula C_6H_{14}. This is indeed the case. The compound corresponding to structure IV is called 2-methylpentane; that corresponding to structure V is called 2,2-dimethylbutane. Both are liquids at room temperature, and their boiling temperatures are 60 °C and 50 °C respectively, quite close to the value for hexane.

▷ Do you find the closeness of the values to that of hexane surprising?

▶ You shouldn't; we noted the dependence of the boiling temperature of hydrocarbons on molecular mass. Although their *structural* formulae differ, the *molecular* formulae, and therefore the molecular masses of all three compounds are identical.

Structure VI shows a branched modification of the structural formula of octane. This is called, 2,2,4-trimethylpentane. It was at one time known as iso-octane, and is important because it is the octane of the well-known 'octane number' of petrol, which we shall discuss shortly. As relatively few organic compounds are discussed in this book, we shall not give a systematic description of the way in which they are named. However, you can perhaps see that the *pentane* in 2,2,4-trimethylpentane comes from the five carbon atoms in the main carbon chain of structure VI, and the *2,2,4-trimethyl* tells us that three methyl (CH_3) groups branch off it, two at the second carbon atom in the chain and the third methyl group at the fourth carbon atom.

Cycloalkanes

If a chain of carbon atoms linked by single bonds is connected up nose to tail to form a ring, hydrogen atoms can then be attached to it until rules 1 and 2 are satisfied. This gives rise to a class of compounds known as **cycloalkanes**. Two examples are given as structures VII and VIII. In structure VII, a chain of six carbon atoms links up nose to tail, so the compound is called cyclohexane. In structure VIII, the tail of a chain of seven carbon atoms links up to the second carbon atom of the chain; this again gives a ring of six carbon atoms, but after rules 1 and 2 have been satisfied, there is a methyl (CH_3) group attached to the ring. The compound is therefore called methylcyclohexane.

Alkenes

In the three types of alkane that we have considered up until now, any two neighbouring carbon atoms have been linked by just a single C—C unit bond. But there exist hydrocarbons in which two adjacent carbon atoms use two of their four unit bonds to bind themselves especially tightly to each other. Structure IX shows ethene, the simplest compound of this type. It is obviously related to ethane (structure II). Both compounds contain two carbon atoms which are linked to each other, but in ethane, each carbon uses just one of its four unit bonds to bind to the other, whereas in ethene, it uses two. Because, in accordance with rule 2, each carbon atom can only form four unit bonds, each carbon atom in ethene binds one less hydrogen than does a carbon atom in ethane, and the ethene molecule, C_2H_4, contains two fewer hydrogen atoms in all. Hydrocarbons like ethene which contain a carbon–carbon double bond are called alkenes. Structure X shows another example, which is derived from hexane. This is called hex-2-ene, because the double bond occurs at the second carbon atom in the chain.

Aromatic hydrocarbons

The structural formulae of **aromatic hydrocarbons** combine features of cycloalkanes and alkenes. Structure XI shows benzene, the simplest example. The characteristic feature of aromatic hydrocarbons is a ring of six carbon

```
H3C  H   H
H3C-C - C - C-CH3
     H   H   H
```

IV

```
    H3C  H
H3C-C - C-CH3
    H3C  H
```

V

```
    H3C  H  CH3
H3C-C - C - C-CH3
    H3C  H   H
```

VI

```
      H  H
  H    \ /    H
   \    C    /
H-C        C-H
  |        |
H-C        C-H
   /    C    \
  H    / \    H
      H   H
```

VII

```
     CH3 H
  H    \ /    H
   \    C    /
H-C        C-H
  |        |
H-C        C-H
   /    C    \
  H    / \    H
      H   H
```

VIII

```
H       H
 \     /
  C = C
 /     \
H       H
```

IX

```
     H   H   H   H
H3C-C = C - C - C-CH3
            H   H
```

X

```
        H
        |
 H      C      H
   \  //  \  /
     C      C
     |      ||
     C      C
   /  \\  /  \
 H      C      H
        |
        H
```

XI

Smaller molecules easier to burn.

$$\begin{array}{ccccc} & & CH_3 & & \\ & & | & & \\ H & C & = C & C & H \\ & | & & \| & \\ H & C & = C & C & H \\ & & | & & \\ & & H & & \end{array}$$

XII

atoms linked by six carbon–carbon bonds; our structural formula displays three of these carbon–carbon bonds as single, and three as double, with the double and single bonds alternating as one moves around the ring. Rules 1 and 2 then permit just one hydrogen to be attached to each carbon atom in the ring, so the molecular formula of the benzene molecule is C_6H_6. A family of aromatic hydrocarbons can then be built up by replacing one or more of these hydrogen atoms with a hydrocarbon group or chain. Structure XII shows toluene, the simplest compound of this type — one of the hydrogen atoms of benzene has been replaced by a methyl (CH_3) group. ■

The two questions below test your understanding of Box 3.1.

Question 3.1 Only two of the five structural formulae shown as structures XIII–XVII represent known hydrocarbons. Select the two, and explain why the remaining three represent compounds that cannot exist.

$$\begin{array}{c} \quad CH_3 \\ \quad | \\ H_3C-C-CH_2-CH_3 \end{array}$$

XIII

$$\begin{array}{c} CH_3 \\ | \\ H_3C-C-H \\ | \\ CH_3 \end{array}$$

XIV

$$\begin{array}{c} \quad\quad H \;\; H \\ \quad\quad | \;\;\; | \\ H_3C-C=C-CH_3 \\ | \\ H \end{array}$$

XV

$$\begin{array}{ccccc} & H & CH_3 & H & \\ H & & C & & H \\ H-C & & & & C-H \\ | & & & & | \\ H-C & & & & C-H \\ H & & C & & H \\ & H & & H & \end{array}$$

XVI

$$\begin{array}{c} \quad H_3C \;\; H \\ \quad\; | \quad\; | \\ H_3C-C=C-H \end{array}$$

XVII

Question 3.2 There are five compounds with the molecular formula C_6H_{14}. Three of them are shown as structures III–V. In structure IV, a methyl (CH_3) group is attached to the second carbon atom in a five-carbon-atom chain; in one of the two missing structures it is attached to the third. In structure V, two methyl groups are attached to the second carbon atom in a four-carbon-atom chain; in the other missing compound, one methyl group is attached to the second, and one to the third.
Draw structural formulae for the two missing compounds, and make approximate estimates of their boiling temperatures.

3.2 Engine knock and octane number

To make the study of engine knock properly scientific, one needs a quantitative measure of the knock resistance of different fuels. This is done by first selecting a hydrocarbon that is an excellent car engine fuel and causes very little engine knock. The compound chosen is 2,2,4-trimethylpentane, which was shown in structure VI, and whose name will from now on be abbreviated to TMP. To this, a knock resistance index of 100 is assigned. Next one chooses a hydrocarbon that readily causes knocking; a suitable candidate is heptane, and to this a knock resistance index of zero is assigned. As TMP contains eight carbon atoms and was once known as iso-octane, the index has long been known as the **octane number**.

Different fuels can now be assigned an octane number by burning them in a test engine whose compression ratio can be varied. The test engine runs at 600 rev. min^{-1}

with an air inlet temperature of 52 °C, and with the engine coolant at 100 °C. The engine knock produced under these conditions is then compared with that of a mixture of heptane and TMP. When the mixture of TMP and heptane is such that knocking in the mixture and in the fuel under test begins at the same compression ratio, the octane number of the fuel is equal to the percentage of TMP by volume in the heptane–TMP mixture.

▷ The engine knock of a petrol under standard engine conditions is identical with that of a mixture of 1 litre of heptane with 4 litres of TMP. What is the octane number of the petrol?

▶ 80; the sum of the volumes of heptane and TMP is 5 litres, and the volume of TMP is four-fifths of this. Four-fifths expressed as a percentage is 80%.

Although we refer here simply to octane number, the figures obtained with an air inlet temperature of 52 °C at 600 rev. min^{-1} are strictly known as **research octane numbers**, abbreviated to RON. Thus the petrol just discussed would be described as 80 RON petrol. RON values are the best measure of fuel performance under mild driving conditions, for example when cruising or gently accelerating. An often quoted alternative is the *motor* octane number (MON). This is determined with the air intake temperature at 38 °C, and an engine rating of 900 rev. min^{-1} In contrast to the Research method, the temperature of the air–fuel mixture prior to combustion is also specified (149 °C). MON values are appropriate under more severe conditions, like motorway driving. However, in this book, we shall use RON values only.

In terms of knock resistance, some fuels are worse than heptane and so have negative octane numbers, while others are better than TMP and have octane numbers above 100. In these cases, the method of RON measurement has to be modified to extend the scale both downwards and upwards, but we shall not go into details here.

3.2.1 Octane number and molecular structure

Table 3.2 lists the octane numbers of some hydrocarbon fuels containing six carbon atoms. One compound has been included from each of the five categories in Box 3.1. Also present are the data on heptane and TMP which define the octane number scale.

▷ How knock-resistant are straight-chain alkanes?

▶ Not very knock-resistant at all; hexane, heptane and octane are the straight-chain alkanes in Table 3.2, and their octane numbers are very low. The straight-chain alkanes become even less knock-resistant as the chain gets longer; octane, for example, is even worse than heptane, and so has a negative RON value.

▷ Are branched-chain alkanes better fuels than their straight-chain counterparts?

▶ Much better; 2,3-dimethylbutane has the same formula as hexane, but has two branches in the chain, and its octane number is much higher. Likewise, there is an enormous improvement in moving from octane to TMP, even though both have the formula C_8H_{18}.

A comparison of the data for hexane, cyclohexane and hex-2-ene suggests that the formation of rings of carbon atoms, and of carbon–carbon double bonds both have a very beneficial effect upon octane number, but the greatest knock-resistance seems to be found when both these features are combined, as in aromatic compounds.

Table 3.2 Research octane numbers (RONs) of some hydrocarbons. First come the compounds heptane and TMP whose degrees of knock-resistance fix the octane number scale; after that, the compounds are arranged in order of ascending RON values.

Hydrocarbon	Structural formula	RON	Type of hydrocarbon
heptane	$H_3C-(CH_2)_5-CH_3$	0	straight-chain alkane
2,2,4-trimethylpentane (iso-octane or TMP)	$\begin{array}{ccccccccc} & & CH_3 & & & & CH_3 & & \\ & & \vert & & & & \vert & & \\ H_3C & - & C & - & CH_2 & - & C & - & CH_3 \\ & & \vert & & & & \vert & & \\ & & CH_3 & & & & H & & \end{array}$	100	branched-chain alkane
octane	$H_3C-(CH_2)_6-CH_3$	−19	straight-chain alkane
hexane	$H_3C-(CH_2)_4-CH_3$	25	straight-chain alkane
2,3-dimethylbutane	$\begin{array}{ccccccc} & & H_3C & & CH_3 & & \\ & & \vert & & \vert & & \\ H_3C & - & C & - & C & - & CH_3 \\ & & \vert & & \vert & & \\ & & H & & H & & \end{array}$	72	branched-chain alkane
cyclohexane	$\begin{array}{ccc} & H_2 & \\ & C & \\ H_2C & & CH_2 \\ \vert & & \vert \\ H_2C & & CH_2 \\ & C & \\ & H_2 & \end{array}$	83	cycloalkane
hex-2-ene	$H_3C-CH=CH-(CH_2)_2-CH_3$	92	alkene
benzene	$\begin{array}{ccc} & H & \\ & C & \\ HC & & CH \\ \vert & & \Vert \\ HC & & CH \\ & C & \\ & H & \end{array}$	106	aromatic hydrocarbon

Activity 3.1

Petrols consist mainly of a mixture of straight-chain alkanes, branched-chain alkanes, cycloalkanes, alkenes and aromatic hydrocarbons. Suppose that a type of chemical reaction could be devised which converted one of the categories into another. To judge by Table 3.2 and Section 3.2.1, what particular type of conversion would bring about the greatest improvement in the knock-resistance of the petrol? How would you expect the conversion to affect (a) the compression ratio at which the petrol starts knocking and (b) the composition of the heptane–TMP mixture which matches this petrol in the RON measurement?

3.3 Making petrol

Crude oil is chiefly a mixture of alkanes, cycloalkanes and aromatic hydrocarbons. When petrol is made from crude oil, the first important step is fractional distillation (Figure 3.3). This separates the oil into fractions with different boiling temperature ranges. Four fractions appear in Figure 3.3.

▷ Which of the four has the highest boiling temperature range?

▶ Gas oil; it is liquefied within its boiling temperature range in a lower section of the column where, as the caption to Figure 3.3 states, the column temperature is relatively high. The other three are vapours at this stage, because they are above their boiling temperature ranges.

During the induction stroke of Figure 3.1, the fuel in a petrol engine must be volatile enough to be drawn into the cylinder as a mixture of vapour and liquid droplets. But it must not be so volatile that pockets of vapour appear and block the fuel lines, or so that excessive fuel is lost by evaporation. The normal boiling temperature range of a petrol that meets these requirements is 30–190 °C.

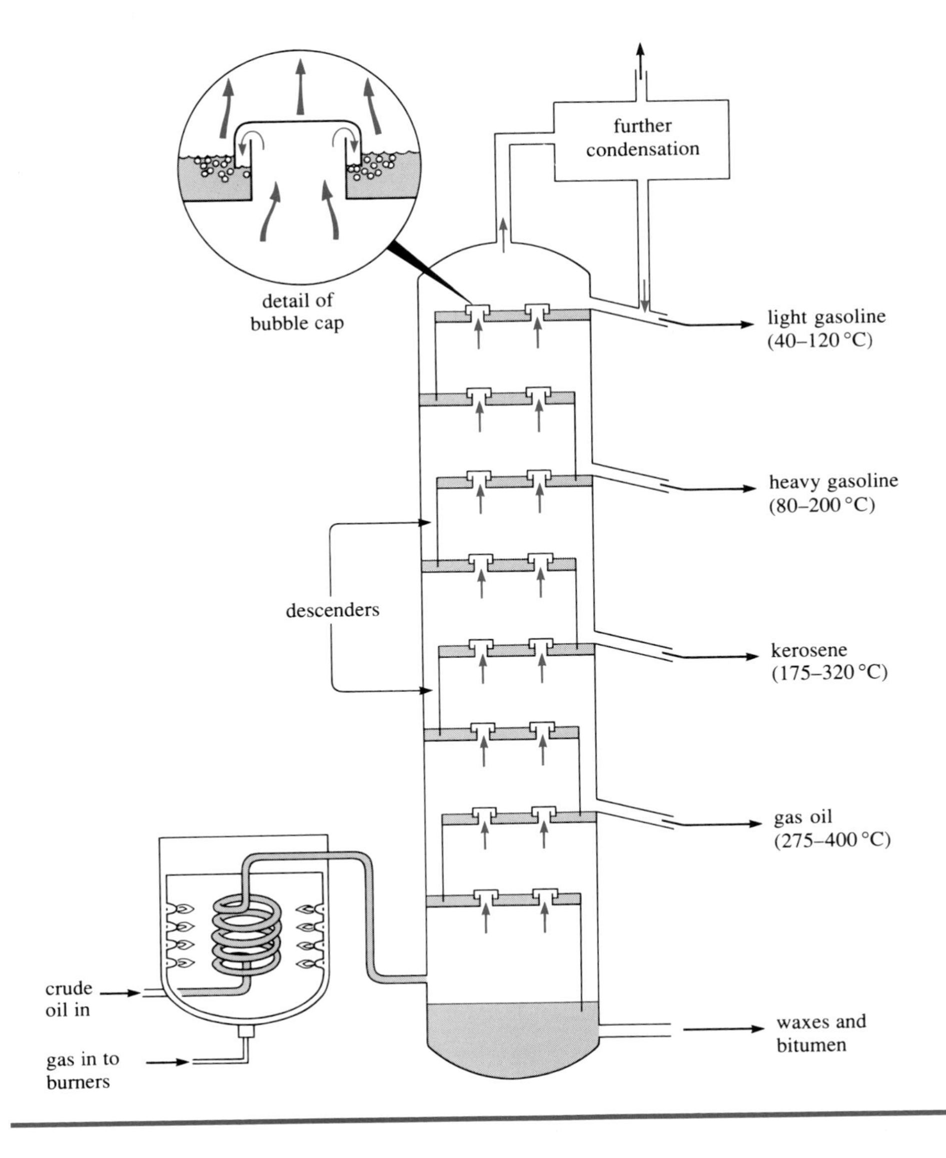

Figure 3.3 In the fractional distillation of crude oil, the oil is heated in a furnace, and its vapour then ascends a tall tower which is broken into sections by perforated trays. Above the perforations are bubble-shaped caps where components of the ascending vapour can condense to a liquid which becomes part of a shallow layer on the tray. Some liquid flows back through the descenders to lower trays where its constituents may be re-evaporated, and rejoin the ascending vapour. The tower may be 50 metres high, and the temperature of the sections decreases as one moves up the column away from the furnace. In the hot, lower sections, only the hydrocarbons with the highest boiling temperatures are removed from the vapour by condensation, but as the vapour ascends, hydrocarbons with progressively lower boiling temperatures are liquefied. Thus on different trays on the column, mixtures of the oil's constituent hydrocarbons with particular ranges of boiling temperature are condensed to a liquid and can be run off. These mixtures are called **fractions**. In the figure just four important fractions are shown: light and heavy gasoline, kerosene and gas oil. In the refining industry, the fractions called gasolines in this book are sometimes called *napthas*.

▷ Which of the fractions in Figure 3.3 might therefore be used for petrol?

▶ The light and heavy gasoline fractions cover this boiling temperature range.

The two gasoline fractions can be mixed in different proportions, or used individually. In either case, the result is known in the USA as straight-run gasoline or in the UK, as straight-run petrol. But however it is put together, **straight-run petrol** turns out to be a quite unsuitable fuel for the modern motor car. This is because its octane number is typically about 60, but the modern petrol engine works at compression ratios that require fuel with an octane number greater than 90. How then can the octane number of straight-run petrol be increased?

3.3.1 Improving straight-run petrol

To answer the question at the end of the preceding section, one needs to know more about the content of straight-run petrol, and of the gasoline fractions from which it is made.

▷ Given that the fractions in Figure 3.3 consist almost entirely of hydrocarbons, will the numbers of carbon atoms in the molecules, and therefore the molecular masses, be greater in gas oil, or in the gasoline fractions?

▶ The number of carbon atoms, and hence the molecular masses, will be greater in the gas oil fraction. This is because the boiling temperature range of gas oil is higher than that of the gasoline fractions, and in a family of hydrocarbons, boiling temperature increases with molecular mass, as seen in Table 3.1.

In fact, the hydrocarbon molecules in gas oil typically contain 12–18 carbon atoms; those in the gasoline fractions, 5–10 carbon atoms.

In straight-run petrols, the 5–10 carbon atom hydrocarbons consist chiefly of straight-chain alkanes. The next most abundant are usually cycloalkanes, followed by branched-chain alkanes and aromatics. The straight-chain alkanes are mainly responsible for the low octane number. Consequently, the petrol will be greatly improved if it undergoes chemical reactions that convert low-octane substances such as straight-chain alkanes into high-octane products like aromatics and branched-chain alkanes. The most important way of doing this is called **catalytic reforming**.

3.3.2 Catalytic reforming

In the most modern reforming plants, straight-run petrol vapour is brought to a pressure of 4–10 atmospheres and a temperature of perhaps 500 °C, in the presence of metallic platinum and rhenium. The metal is very thinly spread over a supporting material, usually aluminium oxide, Al_2O_3; this arrangement maximizes the exposure of the hydrocarbons to the metal surface. The platinum–rhenium acts as a **catalyst**; although unchanged itself, it speeds up chemical reactions that are almost, or completely imperceptible when it is absent. The reactions can be likened to a cart, immobilized on a slope because its wheels and axles are rusty. The slope favours downward motion but the motion is immeasurably slow because of the rust. However, if one oils the wheels, the motion is speeded up and becomes perceptible. For a chemical reaction, the material which emulates the oil is known as a catalyst: a substance which can speed up the reaction without undergoing changes itself. Typical catalysed reactions in this case are the conversions of straight-chain alkanes to cycloalkanes, and then to aromatics:

$$CH_3—(CH_2)_4—CH_3(g) \longrightarrow C_6H_{12}\text{ (cyclohexane: ring of six }CH_2\text{ groups)}(g) + H_2(g) \quad (3.2)$$

$$C_6H_{12}\text{ (cyclohexane)}(g) \longrightarrow C_6H_6\text{ (benzene: ring of six }CH\text{ groups)}(g) + 3H_2(g) \quad (3.3)$$

Another important change during catalytic reforming is the conversion of straight-chain to branched-chain alkanes:

$$CH_3—(CH_2)_4—CH_3(g) \longrightarrow CH_3—\underset{H}{\overset{CH_3}{C}}—(CH_2)_2—CH_3(g) \quad (3.4)$$

As the examples we have given suggest, reforming is characterized by little or no change in the number of carbon atoms in the hydrocarbon.

Catalytic reforming is usually carried out in the presence of hydrogen gas to protect the catalyst from impurities such as sulphur. Sulphur is carried away as the gas hydrogen sulphide, H_2S. A typical starting material will have an octane number of 60, and contain 60% alkanes, 25% cycloalkanes, and 15% aromatics. This is converted to a product of octane number 98, containing perhaps 32% alkanes which are mainly branched, 2% cycloalkanes, and 66% aromatics.

3.3.3 Catalytic cracking

Catalytic reforming improves the quality of straight-run petrol, but does not deal with the problem of its quantity. Of the different fractions obtained by the distillation of crude oil, light and heavy gasoline are the ones most in demand, yet when the world's most abundant crude oils, such as those of Saudi Arabia or Kuwait, are distilled, the gasoline fractions amount to less than 20% of its volume. But this proportion can be greatly increased by taking one of the less valuable, higher boiling temperature fractions such as gas oil, and converting the constituents into high-octane hydrocarbons with boiling temperatures in the gasoline range. This role is performed by **catalytic cracking**.

The higher boiling temperature range of gas oils is a mark of the larger molecules and higher molecular masses of the hydrocarbon components; a typical constituent will contain 12–18 carbon atoms.

▷ What will be the formula of a straight-chain alkane with 12 carbon atoms?

▶ $C_{12}H_{26}$; the general alkane formula is C_nH_{2n+2}, and here $n = 12$.

Cracking is the name given to a chemical process in which such long-chain alkanes break up into two smaller molecules. Catalytic cracking is conducted at about 525 °C in the presence of a zeolite catalyst. Zeolites are porous solids composed mainly of silicon, aluminium and oxygen atoms. In a typical step, the breaking of the carbon chain leads to the formation of an alkane and an alkene; for example:

$$CH_3(CH_2)_{10}CH_3(g) \longrightarrow CH_3(CH_2)_6CH_3(g) + CH_2{=}CHCH_2CH_3(g) \quad (3.5)$$

However, the chain may break at different points, and the alkane that is formed may then be cracked in its turn and perhaps converted into a more branched form. Thus the ultimate products of the cracking of the $C_{12}H_{26}$ molecule might, under the chosen conditions, be a branched alkane containing five carbon atoms, such as $CH_3CH_2CH(CH_3)_2$, and two alkene molecules, $CH_2{=}CHCH_3$ and $CH_2{=}CHCH_2CH_3$. With only three and four carbon atoms respectively, the two alkenes have low molecular masses and low boiling temperatures; they remain gaseous at a temperature at which the alkane and similar compounds can be separated by condensation, and used in petrol. Because of its branched-chain structure, the alkane will have a favourable octane number as well. At the same time, the other products—the volatile alkenes—are fed into other parts of the petrochemical industry. In other cases, the carbon chain may break to give an alkene which contains five or six carbon atoms, has a higher boiling temperature, and can therefore be used in petrol; for example:

$$CH_3(CH_2)_{10}CH_3(g) \longrightarrow CH_3(CH_2)_4CH_3(g) + CH_2{=}CH(CH_2)_3CH_3(g) \qquad (3.6)$$

As Table 3.2 suggests, the octane numbers of alkenes are quite high.

As the demand for motor car fuels grew between 1910 and 1940, it was the exploitation of cracking and other processes that enabled the oil industry to increase the supply of petrol, and to overcome the problem of the low percentage of gasoline in crude oil.

3.3.4 Petrol and lead alkyls

The high-octane product of catalytic cracking can be blended with catalytically reformed gasoline fractions to give a high-octane petrol (Figure 3.4). During the 1970s, the average octane number of the product was about 89 RON. But most drivers used 4-star petrol, which is characterized by an octane number of at least 97 RON, so further upgrading was necessary. This was done by adding lead alkyls (Figure 3.5).

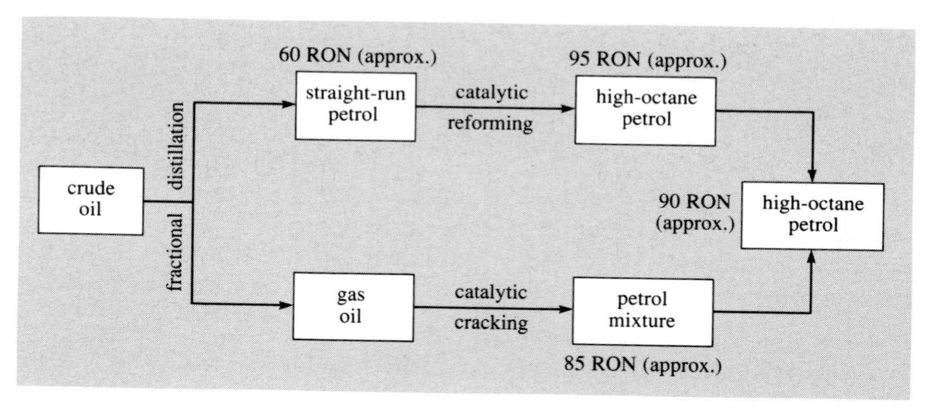

Figure 3.4 How the fractional distillation of crude oil, catalytic reforming and catalytic cracking can be combined to produce high-octane petrol.

Structure XVIII shows the structural formula of tetraethyl lead. It has often been said that it 'mimics' a hydrocarbon.

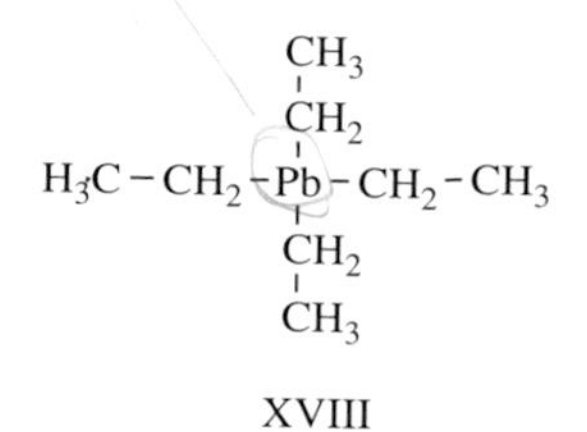

▷ How is this?

▶ It is identical with a hydrocarbon except for the central atom where lead replaces carbon. Moreover, the substituted lead atom, like the carbon it replaces, forms four single bonds to other carbon atoms.

Tetraethyl lead is a colourless liquid with a boiling temperature of about 200 °C. Because of its hydrocarbon-like structure, hydrocarbon molecules accept it as a fellow molecule, so it dissolves quite easily in the hydrocarbon mixture we call petrol. In an

Figure 3.5 Few scientific inventions receive both extensive worldwide acclaim and a subsequent hostile rejection. Thomas Midgely (1889–1944) made two of them. It was clear even before the First World War that the problem of engine knock would obstruct the development of more efficient, high-compression engines. In 1917, Midgely began testing the effect of various additives on the knock-resistance of petrol at the Dayton Engineering Laboratories Company in Ohio. The tellurium compound, diethyl telluride, $Te(CH_2CH_3)_2$, looked a winner at one stage, but it graced all those who handled it with a repulsive, garlic body odour, and was dropped. Some 30 000 different substances had been tried before the crucial morning of 9 December 1921. On that day, Midgely and his team stood beside a test engine which, under the influence of full-blooded engine knock, was labouring like a miniature stamp mill. A small amount of the compound tetraethyl lead, $Pb(C_2H_5)_4$, was injected into the fuel line, and the staccato roar faded dramatically to a smooth and musical purr. In 1937, the *New York Times* remarked that this discovery 'adds fifty times as much horsepower annually to American civilization as that which will be supplied by the Boulder Dam'. The second of Midgely's ultimately ill-fated discoveries was made in the mid-1930s, when he developed chlorofluorocarbon refrigerants (CFCs), which have now been banned in many countries because of their harmful effect on the ozone layer. In 1940, Midgely contracted polio which left him paralysed. He endured his illness with courage and fortitude, but the unforeseen drawbacks of his inventions were apparent to the last: in November 1944, he accidentally strangled himself in a harness he had devised to help himself to get out of bed.

engine that burns petrol to which tetraethyl lead has been added, it inhibits, in a way still not fully understood, the chemical reactions that occur before the arrival of the flame front and cause engine knock.

This use of tetraethyl lead, however, is not without side-effects. When it burns, along with the petrol to which it has been added, it forms solid lead oxides, such as PbO, that are deposited on the sparking plugs and cylinder walls, thereby shortening the engine life. To prevent this, compounds such as 1,2-dibromoethane (structure XIX) are added to the petrol as well. The combustion of tetraethyl lead then produces compounds such as lead dibromide, $PbBr_2$. Unlike the oxides, lead dibromide is volatile at the temperature of combustion, and is swept out of the cylinder with the exhaust gases. In the cooler parts of the exhaust system, it condenses to fine particles, in which form it emerges into the atmosphere. Some hydrogen bromide gas, HBr, is produced at the same time, however. Like HCl (Box 1.2) this is acidic, and increases corrosion in the exhaust system.

```
    H  H
    |  |
Br—C—C—Br
    |  |
    H  H
```

XIX

On the other hand, some of the side-effects of tetraethyl lead are beneficial, most notably the reduction of wear at the surfaces of contact between engine valve heads and valve seats (Figure 3.6).

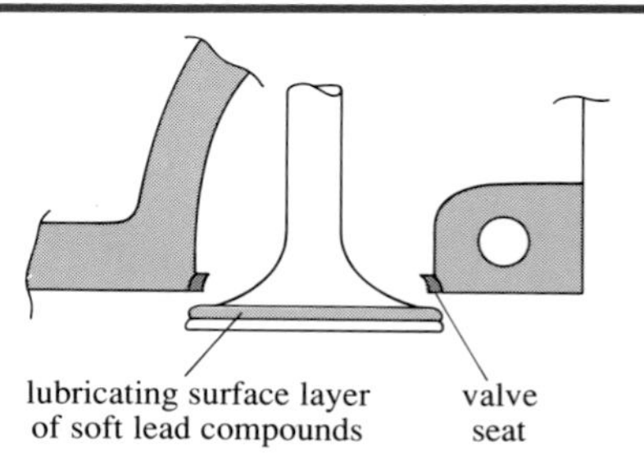

Figure 3.6 The two valves, shown at the top of the cylinder head in Figure 3.1, each open and shut once in the 4-stroke cycle, one admitting the air–fuel mixture, and the other expelling the combustion products. Leaded petrol leaves a deposit of condensed, solid lead compounds on the surface at the cool side of the valve head which cushions the continual impact between valve head and valve seat, thereby restricting wear which might otherwise lead to a loss of compression.

The crucial point, however, is that tetraethyl lead, along with the related compound, tetramethyl lead (structure XX), provides a cheap and effective way of boosting the octane rating of petrol. In the 1960s and 1970s, the addition of tetraethyl lead at a level which put about 0.8 g of lead into each litre of petrol* was used to boost the octane number of the unleaded, refined product from about 90 to 97.

$$\begin{array}{c} CH_3 \\ | \\ H_3C-Pb-CH_3 \\ | \\ CH_3 \end{array}$$

XX

3.4 Lead in petrol; the options

This chapter has explained, in scientific language, some of the technical and material constraints which prompted the use of leaded petrol. You are now asked to clarify and summarize its main points in your own mind by doing Activity 3.2. Here, for the first time, you are being asked explicitly to communicate your views in written form. To make things easier, the exercise has been broken down into three smaller parts. Deal with each part in turn, first writing out notes on a possible answer from what you can remember, then going back to the text to check and fill out your recollections. Finally write your answer out in full rather than in note form; this will provide useful practice for TMAs.

Activity 3.2 *You should spend up to 30 minutes on this activity.*

Explain in less than 200 words why lead compounds do not play an irreplaceable role in raising the octane number of petrol to a figure at which car engines can operate. Structure your answer in the following way:

(i) Describe the principal methods used to raise the octane number of petrol to the 4-star level, and discuss the importance of the part played by lead additives in this process.

(ii) Propose one method by which the octane number might be maintained at the 4-star level after lead has been completely eliminated from petrol.

(iii) Explain how car engines might be modified to make acceptable the loss of octane number which might follow the removal of lead from petrol. What effect would these changes have on power and petrol consumption?

When you have completed the task, compare your response with the answer on pp. 83–84.

The alternatives to leaded petrol were widely discussed in the 1960s and 1970s as concern grew over environmental lead. However, because all of them involved somebody in additional expense or inconvenience, maintenance of the status quo—the con-

* Note that Extract 2.1, examined in Activity 2.3, gives concentrations which are 1 000 times too small.

tinued use of lead alkyls—retained some attractions. This is therefore included in the five options below which provide a framework within which various ways forward can be discussed:

1 Continue using lead alkyls in petrol at existing concentrations.

2 Fit filters to exhaust systems which will remove the lead. Obviously such filters will cost money.

3 Eliminate or reduce the amount of lead in petrol, and accommodate the loss of the octane number by modifying car engines, notably by reducing engine compression ratios. Engine efficiency is lowered, and because the combustion of the fuel is less complete, the proportion of poisonous carbon monoxide and unburnt hydrocarbons in the exhaust gases is increased.

4 Eliminate or reduce the amount of lead in petrol, and restore the octane number by more intensive refining. This might, for example, involve prolonging catalytic reforming, or carrying it out under more carefully controlled conditions, so that the proportion of aromatics is increased. Such options were considered in Activity 3.2. They are not without drawbacks. For example, if aromatic hydrocarbons fail to burn in the engine cylinders, their 6-carbon rings may instead join up to form larger molecules called **polynuclear aromatic hydrocarbons (PAHs)**. An example is chrysene (Figure 3.7). Increased levels of such substances are also formed in the more intensive reforming processes which are used to make highly aromatic petrol, so such petrols contain more of them prior to burning. Now polynuclear aromatic hydrocarbons are carcinogenic (cancer-causing) chemicals. This is also true of benzene itself. Thus if the proportion of aromatics in petrol is increased, a more carcinogenic urban atmosphere may be the result. In addition, capital costs will be incurred in modifying refineries, and more energy will be needed to prolong the refining. Moreover, in the octane region above 90, refining is subject to diminishing returns—further increases in octane number are only achieved at the cost of a reduction in the amount of petrol obtained.

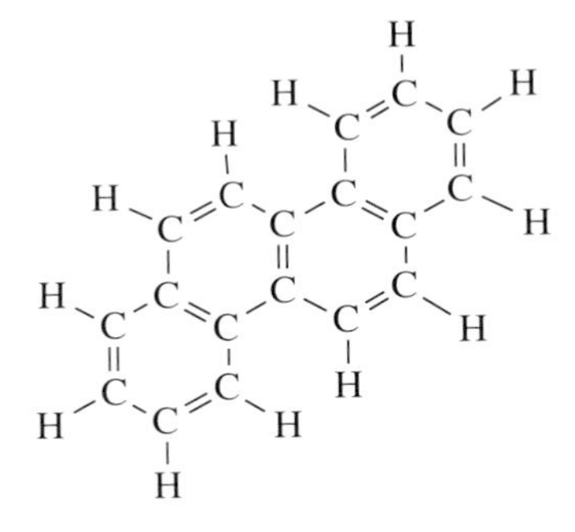

Figure 3.7 Chrysene is an example of a polynuclear aromatic hydrocarbon. It consists of four 6-carbon aromatic rings sharing common sides. The molecular formula is $C_{18}H_{12}$, and the compound is highly carcinogenic. The higher the proportion of aromatics in a petrol, the higher the percentage of such compounds in the exhaust stream tends to be.

5 Eliminate or reduce the amount of lead in petrol, and find other non-hydrocarbon alternatives which will restore the octane number. One kind of alternative is a class of substances known as **oxygenates**, which are essentially hydrocarbons with an oxygen atom inserted at some point in the molecule. This type of solution is not new. One suitable oxygenate is ethanol (alcohol in common parlance) and during the 1930s, one company took advantage of its favourable effect on octane number by marketing a petrol containing 15–16% ethanol (Figure 3.8). The structural formulae of the four most suitable oxygenate components are shown in Table 3.3 along with their names and research octane numbers.

Because of their high octane numbers, these compounds, or mixtures of them, can be dissolved in petrol to increase its octane rating. The high RON value of MTBE makes it especially effective in doing this, and because the oxygenates are much less toxic than lead alkyls they are environmentally more acceptable. But oxygenates must be added in much greater amounts than lead alkyls to produce the same effect—as much as 5–20% by volume. This option is therefore not so much one of finding alternative additives as of changing the composition of petrol. It requires amounts of oxygenates in the millions of tonnes per year range, and calls for considerable new investment in chemical plant.

In the UK during the 1980s, options 1 and 2 were definitely ruled out, and a decision was taken to pursue some combination of options 3, 4 and 5 by first reducing the legal upper limit of lead in petrol to $0.15\,g\,l^{-1}$, and then, through the 1990s, progressively phasing out lead additives altogether. A well organized environmental campaign had a considerable influence on this decision. Why was the outcome favourable to it? In the next two chapters we shall try to answer this question.

Figure 3.8 Illustrations from the magazine *Speed*: (top) July 1937; (bottom) May 1936. In the 1930s, the Cleveland Oil Company marketed an alcohol–petrol blend containing 15–16% alcohol. The octane number was about 75 at a time when most conventional petrols had values of about 73.

Table 3.3 Structural formulae and octane numbers of some oxygenates that can be used to increase the octane number of petrol.

Structural formula	Name	RON
$CH_3{-}O{-}H$	methanol	112
$CH_3{-}CH_2{-}O{-}H$	ethanol	111
$CH_3{-}C(CH_3)_2{-}O{-}H$ (central C bonded to CH_3 above and below)	2-methyl-2-propanol (commonly called tertiary butyl alcohol, TBA)	113
$CH_3{-}C(CH_3)_2{-}O{-}CH_3$ (central C bonded to CH_3 above and below)	methyl 2-methyl-2-propyl ether (commonly called methyl tertiary butyl ether, MTBE)	117

Summary of Chapter 3

1 In the four-stroke petrol engine, a mixture of air and hydrocarbons is drawn into the engine cylinders, compressed and ignited by a spark. The gaseous products of the combustion reaction expand, pushing the piston back, before being expelled from the cylinder in an exhaust stroke.

2 Engine knock results from premature explosive chemical reactions which occur in front of the advancing flame front during the ignition stroke. Unfortunately, it occurs more readily at higher compression ratios, where engine power and fuel economy are greater.

3 The research octane number (RON) is a measure of the resistance of a fuel to engine knock. It is the percentage of TMP by volume in a heptane–TMP mixture which has the same susceptibility to knocking as the fuel.

4 Straight-chain alkanes have very low RON values. The values for branched-chain alkanes, cycloalkanes and alkenes are much higher, and those for aromatic hydrocarbons are higher still.

5 When crude oil is fractionally distilled, the light and heavy gasoline fractions in the petrol boiling temperature range are usually only about 20% of the oil volume, and have low RON values of about 60–70. They are called straight-run petrols.

6 Catalytic reforming of straight-run petrol increases the RON value to 95–100, principally by destroying straight-chain alkanes and creating aromatics.

7 Catalytic cracking breaks up the hydrocarbon molecules in gas oil and other higher boiling temperature fractions, producing smaller high-octane hydrocarbon molecules with boiling temperatures in the petrol range. The product can then be blended with catalytically reformed straight-run petrol, greatly increasing the volume of petrol obtainable from crude oil.

8 In the 1960s and 1970s, petrol made in this way was 85–90 RON. The final lift to 97 RON 4-star petrol was provided by addition of tetraethyl and tetramethyl lead whose combustion products were emitted through the exhaust.

9 Possible alleviative measures include the fitting of lead filters, lower compression ratios, more intensive chemical refining of petrol and substitution of high-volume oxygenates for the low-volume lead alkyls.

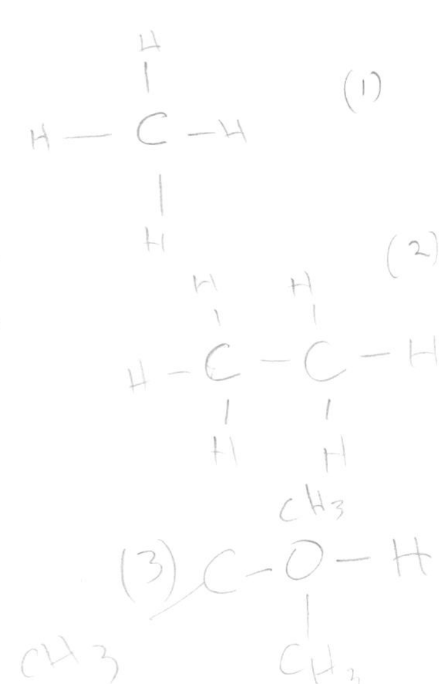

Question 3.3 At the beginning of Box 3.1, the structural formulae of hydrocarbons were said to conform to the following rules:

1 Each hydrogen atom forms just one bond which must link it to a carbon atom.

2 Each carbon atom forms four bonds, any one of which may be linked to either a hydrogen atom, or another carbon atom.

Modify these two rules and add a new rule, rule 3, to make a new set to which structural formulae of both hydrocarbons, *and* the oxygenates shown in Table 3.3 conform.

Question 3.4 Arrange the three compounds whose structural formulae, (i)–(iii), are given below, in order of decreasing octane number.

```
            CH3
            |
(i)   CH3—C—(CH2)3—CH3
            |
            H
```

(ii) $CH_3—(CH_2)_5—CH_3$

```
          H
          C       CH3
      HC ⁄  ＼C ⁄
(iii) ||       |
      HC ＼  ⁄C ＼
          C       CH3
          H
```

Question 3.5 Classify each of the four compounds whose structural formulae, (i)–(iv), are given below into one of the following categories (a)–(d):

(a) a desirable component of high-octane petrol;

(b) an undesirable component of high-octane petrol which could be transformed into a desirable component by catalytic reforming;

(c) an undesirable component of high-octane petrol which could be transformed into a desirable component by catalytic cracking;

(d) none of (a)–(c).

(i) $CH_3—(CH_2)_6—CH_3$ (ii) $CH_3—(CH_2)_{18}—CH_3$

```
          H3C  CH3                     H
           |    |                      |
(iii) H3C—C————C—CH3       (iv)   H—C—H
           |    |                      |
          H3C  CH3                     H
```

4 Generating concern about leaded petrol: the scientific foundations

Like other well organized environmental campaigns, such as that against nuclear power, the attack on leaded petrol combined science with psychology and politics. Whenever people feel themselves at risk from some new kind of pollution, somebody argues that this shows that you cannot trust 'the experts'. Whether this is true or not, sensible environmental campaigners know that their first priority must be to either find or install among the experts, people who are sympathetic to their point of view. Once this has been done, the campaign will have access to authoritative scientific results and arguments that can be used to further its case. The most suitable material of this sort is the kind that generates alarm, fear or depression. Newspapers, television and radio must then be persuaded to transmit the information and its associated psychological unease to the general public. Little persuasion will be necessary if the unease is substantial, because such items make good copy. Once the campaign has gained wide publicity, politicians are forced to take notice, and some may even see advantages in supporting it. A debate begins, opponents are forced to expose themselves, and they can then be openly engaged in argument. This generates even more publicity. Now there is a real possibility that the campaign may achieve its aims through legislation or a change of government policy. But persistence will be necessary, and this requires a continual flow of fresh scientific and technical arguments.

This chapter describes five different categories of scientific work and scientific argument that had particular influence in the debate about environmental lead. In each case, the importance of the issue is assessed by quotations taken directly from the 9th Report of the Royal Commission on Environmental Pollution (1983). This report was the direct cause of the legislation that set in motion the gradual elimination of leaded petrol from the UK.

4.1 The safety margin in blood lead concentrations

The first argument is a very simple one. In Section 2.3 you were shown how blood lead concentrations could be determined by atomic absorption spectrometry (AAS). In about 1980, the average value in UK inner cities was about 13 $\mu g\,dl^{-1}$, and nearly 10% of people in the tested sample had about 20 $\mu g\,dl^{-1}$.

▷ What fraction of the value at which a male lead worker must be suspended from work in the UK, is 20 $\mu g\,dl^{-1}$?

▶ Two-sevenths; suspension is required at 70 $\mu g\,dl^{-1}$ and above (Section 2.4.1).

Now as Table 2.1 shows, symptoms such as colic and frank anaemia, have sometimes been observed close to or, in children, even below this legislative threshold. This is alarming; it appears that in the early 1980s, large numbers of British people were constantly exposed to lead doses about one-quarter of those at which clear symptoms of lead poisoning sometimes appear. There is even the suggestion of emotion in the Royal Commission's pronouncement:

> *The average blood lead concentration in the population is about one-quarter of that at which symptoms of frank poisoning may occasionally occur. We find this disturbing. We do not know of any other toxic substance which is both so widely distributed in human and animal populations and present at concentrations greater than even one-tenth of those at which frank symptoms may occur. We consider this reason enough to seek to reduce the exposure of the general population to lead...* (para. 5.23)

4.2 The accumulation of lead in the global environment

The blood lead concentrations that we have just discussed are about $10\,\mu g\,dl^{-1}$, which is about 1 g of lead in 10 tonnes of blood. Such figures bring home to us the sensitivity of atomic absorption spectrometry, the technique used to obtain them (Section 2.3). But other evidence used to attack leaded petrol included lead concentrations in different layers of Arctic snow. Here the lead levels are about one-ten-thousandth of those in blood—about 1 g of lead in 100 000 tonnes of ice. At the time when the measurements were made, these concentrations were far too small to be determined by AAS, and a different technique had to be used. This is called isotope dilution analysis, and to understand it, you must be familiar with the concept of isotopes and mass spectrometry. If you need to revise this subject, read Box 4.1; if not, pass on to Section 4.2.1.

Box 4.1 Isotopes and the mass spectrometer

4.1

The atoms in a natural sample of lead do not all have the same mass. At the heart of any atom is its nucleus where nearly all the mass resides. The nucleus contains two kinds of particle, protons which carry a single positive charge, and neutrons which carry no charge whatsoever. Protons and neutrons have the same mass. Consequently, the sum of the numbers of protons and neutrons is an indication of the mass of the atom, and it is known as the **mass number**. To be a lead atom, an atom must have 82 protons in its nucleus; that is, all lead atoms contain 82 protons. But some atoms in natural lead have 122 neutrons, some 124, some 125 and some 126.

▷ What are the mass numbers of the four different types of lead atom?

▶ 204, 206, 207 and 208 respectively; the sums of the numbers of protons and neutrons in their nuclei.

These four different types of lead atom which differ in mass, and in number of neutrons, are known as **isotopes** of lead. To distinguish them from each other, the mass number is written to the top left of the chemical symbol. Thus the isotopes with mass numbers 207 and 208 are written ^{207}Pb and ^{208}Pb respectively. Their mass differences do not prevent them from behaving identically in chemical reactions: all four isotopes, for example, form lead oxide, PbO, when heated in air.

A mass spectrometer can sort the atoms of an element into the different isotopes. The essential idea is summarized in Figure 4.1. The four lead isotopes emerge along four distinct

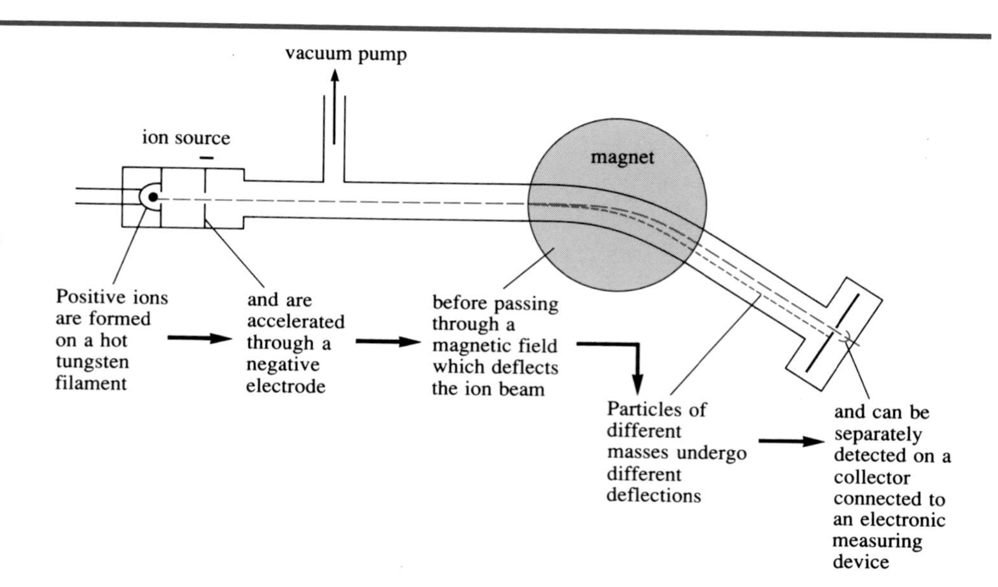

Figure 4.1 The principles and components of a mass spectrometer; in the so-called thermal ion emission source on the left, a suitable lead compound is heated to a very high temperature on a tungsten metal filament in a high vacuum. At these temperatures, the lead compound is broken down into lead atoms which then lose an electron and form individual Pb^+ ions. The charge allows the ions to be accelerated in an electric field. They are then passed between the poles of a magnet which provides a magnetic field at right angles to the ion path. The charged particles are deflected by the magnetic field, but the more massive particles are deflected least, so the ions $^{204}Pb^+$, $^{206}Pb^+$, $^{207}Pb^+$ and $^{208}Pb^+$ separate into four different trajectories. The four different kinds of ion can be collected separately and used to produce an electric current whose size is a measure of their abundance. Notice that under the conditions that exist in the mass spectrometer, the ions are singly charged, whereas in the chemical reactions discussed earlier, the ion of lead is taken to be Pb^{2+}.

pathways in the form of singly-charged positive ions, and just two of these pathways are shown in the figure. This separation allows the four different kinds of ion to be separately collected and recorded as a *mass spectrum*. Figure 4.2 shows such a recording; there is a peak for each isotope, and the peak heights are proportional to the abundances of the different isotopes in the lead sample.

▷ What are the most common and least common isotopes in naturally occurring lead?

▶ ^{208}Pb is the most common and ^{204}Pb the least common; the lines for these isotopes have the largest and smallest heights respectively.

Figure 4.2 also provides quantitative information. For example, the peak heights for ^{208}Pb and ^{207}Pb are in the ratio 2.5 : 1. This tells us that there are about five ^{208}Pb atoms for every two ^{207}Pb atoms in this sample of naturally occurring lead.

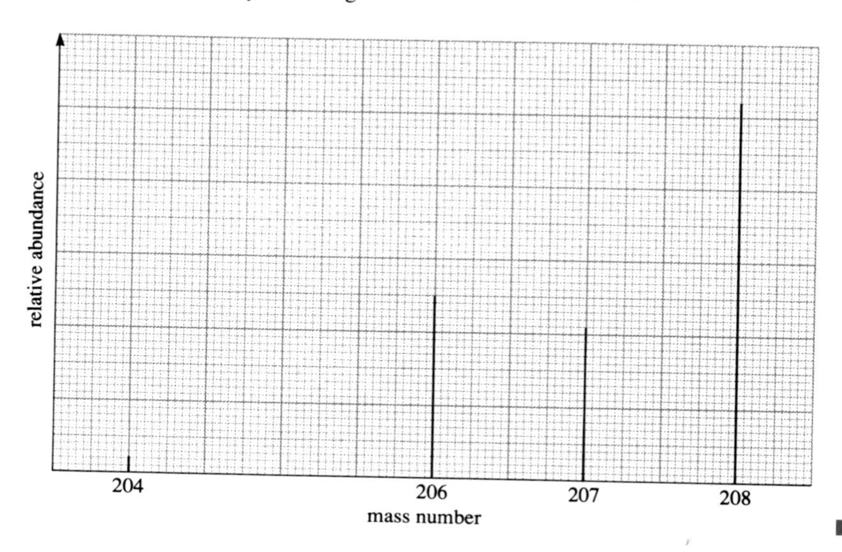

Figure 4.2 The mass spectrum of a sample of naturally occurring lead. The electric current provided by the ions of each isotope is plotted out as a vertical line whose height is proportional to the abundance of the isotope in the lead sample. Because of this proportionality, the vertical axis is simply labelled *relative abundance*, a quantity without units.

■

4.2.1 Isotope dilution analysis of lead in ice

Let us first describe the problem. We have an ice sample with a mass of about 20 kg. It contains as little as 0.2 μg of lead. When the ice melts this tiny amount of lead will be dispersed in about 20 litres of water. How can such a small concentration be measured?

The lead is present in the melted ice as $Pb^{2+}(aq)$, or as sparingly soluble particles which respond like $Pb^{2+}(aq)$ to the chemicals we shall use. Because it is so dispersed, it must be obtained in a more compact or concentrated form if the analysis is to be accurate. Normally, a good way of doing this would be to collect the dispersed lead ions together in a solid which is almost insoluble in water. For example, one could add a source of sulphide ions, $S^{2-}(aq)$, to the melted ice. Now aqueous lead ions and aqueous sulphide ions cannot exist together at appreciable concentrations—this is another way of saying that lead sulphide is almost insoluble in water. Solid lead sulphide is therefore precipitated:

$$Pb^{2+}(aq) + S^{2-}(aq) \longrightarrow PbS(s) \qquad (4.1)$$

The solid contains the lead in a much more compact form, and could then be collected by running the melted ice through a filter.

However, even lead sulphide is not completely insoluble in water, and after it has been precipitated, very small amounts of lead will remain in the solution. Normally, this is insignificant, but in this particular case, the lead concentration was so low to begin with, that the amount left in the solution will be a very substantial proportion of the total—we say that the precipitation is incomplete. Moreover, the amount of precipitate will be so tiny that it will be hard to collect.

These difficulties can be overcome by the technique of **isotope dilution analysis**. We can assume that the lead isotopes in the ice have their normal abundances. Measurements made with the mass spectrometer on many widely dispersed samples show that when this is the case, about 52% of the mass of lead atoms will be ^{208}Pb, and about 21% will be ^{207}Pb, the mass ratio $^{208}Pb/^{207}Pb$ then being about 2.5. Suppose that the mass of the ^{207}Pb atoms in the 20 kg sample of ice is x. Then the mass of ^{208}Pb atoms in the sample will be $2.5x$. The concentrations of lead are such that x would normally be quoted in micrograms (μg).

Next, a known mass of a lead compound in which all the lead consists of the single isotope ^{208}Pb is dissolved in the solution. Suppose that this puts 20 μg of ^{208}Pb into the 20 kg of melted ice. As this is chemically identical with natural lead, we can assume that it becomes evenly and homogeneously mixed with the lead which is already present.

▷ What is the new mass of ^{208}Pb in the sample?

▶ ($2.5x$ + 20 μg); the existing ^{208}Pb mass is supplemented by 20 μg.

▷ What is the new mass ratio, $^{208}Pb/^{207}Pb$, after this supplement?

▶ ($2.5x$ + 20 μg)/x; this is the mass of ^{208}Pb divided by the mass of ^{207}Pb.

We now precipitate lead sulphide as previously described. Our addition of the ^{208}Pb compound has greatly increased the amount of lead in the solution, so this precipitate is now much bulkier, and therefore easier to collect. Even so, a significant proportion of the dissolved lead will remain in the solution. But you will now see that this is unimportant. What *is* important is that we can assume that the relative abundances of ^{207}Pb and ^{208}Pb in the lead sulphide will be in the same ratio as they were in the sample from which the sulphide was precipitated. Such lead sulphide as is obtained is inserted into a thermal emission source mass spectrometer where the lead becomes Pb^{+} ions which are then separated according to their mass (Figure 4.3).

▷ What is the ratio, $^{208}Pb/^{207}Pb$, of the number of atoms in the lead sulphide sample?

▶ The intensities are 450 for ^{208}Pb and 3 for ^{207}Pb, so the ratio is 150.

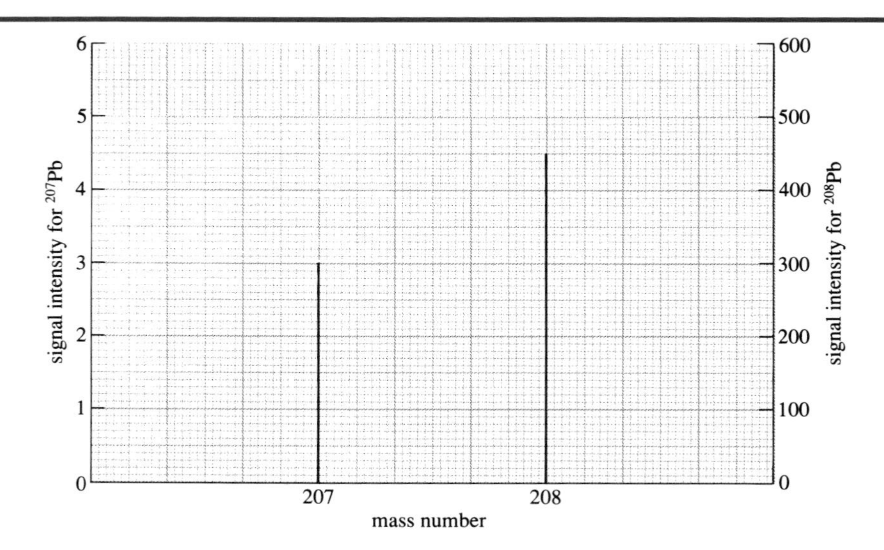

Figure 4.3 The relative abundances of ^{208}Pb and ^{207}Pb in the mass spectrum of a sample of lead sulphide precipitated from melted ice which has been given a ^{208}Pb 'boost'. The relative abundance of the ^{208}Pb signal is recorded on the right-hand vertical axis; that of ^{207}Pb on a different scale on the left-hand vertical axis. The different scales are necessary because the abundances of the two isotopes are so different.

The mass spectrum gives us the ratio of the *numbers* of ^{208}Pb and ^{207}Pb atoms, but the formula we have just calculated is for the ratio of their *masses*. However, the isotopes ^{208}Pb and ^{207}Pb differ in mass number by only 1 unit, so their masses are very similar. Let us assume that they are the same. If we do this, the calculation becomes much simpler, and the error introduced is much smaller than the uncertainties incurred during the experimental measurements. If the ratio of the numbers of ^{208}Pb and ^{207}Pb atoms is 150, and we assume that the masses of the two kinds of atom are identical, then this means that the mass ratio ^{208}Pb/^{207}Pb is 150 as well.

▷ Now calculate the mass, x, of ^{207}Pb in the ice sample.

▶ $$\frac{2.5x + 20\,\mu\text{g}}{x} = 150$$

Multiplying both sides by x:

$$2.5x + 20\,\mu\text{g} = 150x$$

$$\text{Therefore } 20\,\mu\text{g} = 147.5x$$

$$\text{So } x = \frac{20\,\mu\text{g}}{147.5}$$

$$= 0.136\,\mu\text{g}$$

Assuming that the lead isotopes are present at their normal abundances, then as noted already, the mass of ^{207}Pb atoms is about 21% of the total mass of lead.

▷ So what is the total mass of lead in the original ice sample?

▶ 21 µg of ^{207}Pb are present in 100 µg of lead.

So 1 µg of ^{207}Pb is present in 100/21 µg lead and

0.136 µg of ^{207}Pb are present in 0.136 × 100/21 µg lead.

$$\text{Therefore total mass of lead in sample} = 0.136 \times 100/21\,\mu\text{g}$$

$$= 0.648\,\mu\text{g}$$

As the total mass of the sample is 20 kg, the concentration of lead, expressed in micrograms per kilogram of ice is calculated as follows:

$$\text{Concentration of lead} = 0.648\,\mu\text{g}/20\,\text{kg}$$

$$= 0.032\,\mu\text{g}\,\text{kg}^{-1} \text{ (2 significant figs)}$$

The technique is especially valuable because it does not rely on extracting all the lead from the ice sample as PbS(s). All that is necessary is that the ratio $^{208}Pb/^{207}Pb$ in the lead sulphide should accurately reflect the $^{208}Pb/^{207}Pb$ ratio, after this has been supplemented by the known amount of ^{208}Pb.

4.2.2 Isotope dilution analysis of the Arctic ice layers

The climate in the extreme polar regions of the world is arid. The annual layers of snow are usually quite thin, and there is little mixing of any layer with those deposited in previous years. Thus the age of the snows increases with depth, and the different layers can be dated. Dr Clair Patterson (Figure 4.4) used isotope dilution analysis to study the lead concentrations of the differently aged layers in northern Greenland. His results are shown in Figure 4.5. They suggest that the lead concentration in snows deposited from the atmosphere increased enormously between 800 BC and 1965 AD: the detailed figures suggest 200-fold. The increase was especially severe after 1940—a time when one familiar source of atmospheric contamination, the emissions from lead smelters, had been reduced because they were wasting potential production. Patterson's conclusion was simple: the steep increase after 1940 was caused by the huge rise in the consumption of leaded petrol.

Figure 4.4 In 1965, Clair Patterson (1922–) of the California Institute of Technology published a paper which initiated the modern campaign against environmental lead. He has an international reputation in the field of isotope dilution mass spectrometry.

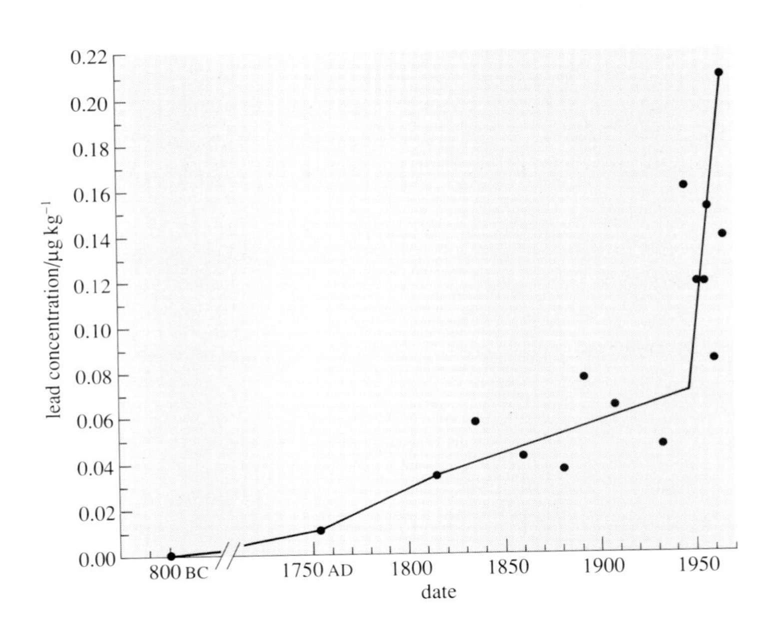

Figure 4.5 The estimated lead concentration of Greenland snows between 800 BC and 1965 AD.

4.2.3 The image of an accumulating global poison

Besides working on the Arctic ice layers, Patterson also made measurements of the lead levels in the bones of Peruvians who had been buried 1 600 years ago, when lead technology in South America was of little or no significance. A typical lead concentration was about 0.1 μg in each gram of dry bone. This is 1/230 of the $23\,\mu g\,g^{-1}$ average obtained in a study of 1 393 people who died in Paris between 1970 and 1975.

At the same time, estimates were made by J. O. Nriagu of the amounts and sources of lead emitted each year into the Earth's atmosphere. The Royal Commission report quoted some of his data which gave total lead emissions in the 1970s of 474 000 tonnes per year. Of this, natural sources such as volcanoes and dust produced by erosion were responsible for only 24 000 tonnes, and the balance of 450 000 tonnes arose from human activities such as lead smelting and the use of leaded petrol. Leaded petrol was especially blameworthy, for this contributed 273 000 tonnes per year, nearly 60% of all emissions. Table 4.1 shows one of Nriagu's more recent analyses covering the emissions of 13 metals in 1983.

Table 4.1 Data on natural and anthropogenic emissions of trace metals to the atmosphere in 1983. Anthropogenic is synonymous with 'man-made', a term proscribed as sexist in these sensitive times.

Trace metal	Emissions			Natural/total
	Anthropogenic	Natural	Total	
		$/10^3$ tonnes		
arsenic	19	12	31	0.39
cadmium	7.6	1.3	8.9	0.15
chromium	30	44	74	0.59
copper	35	28	63	0.44
mercury	3.6	2.5	6.1	0.41
manganese	38	317	355	0.89
molybdenum	3.3	3.0	6.3	0.48
nickel	56	30	86	0.35
lead	332	12	344	0.03
antimony	3.5	2.4	5.9	0.41
selenium	6.3	9.3	15.6	0.60
vanadium	86	28	114	0.25
zinc	132	45	177	0.25

▷ For which two metals are the proportions of anthropogenic emissions the greatest fraction of the total?

▶ Lead (97%) and cadmium (85%). Subtract the figures in the last column from one, and multiply by 100.

▷ For which two metals are the total masses of anthropogenic emissions the greatest?

▶ Lead (332 000 tonnes) and zinc (132 000 tonnes).

By combining figures like these with the work on bones and Arctic ice, Patterson, Nriagu and others conjured up the image of a prehistoric world where natural forces kept lead under proper restraint—a world that was now lost to us because, over the centuries, industry and technology had comprehensively polluted both the planet and its inhabitants with lead. When carrying science into the political arena, it is important to drive results home with forceful pictures, and these were not wanting. Figure 4.6

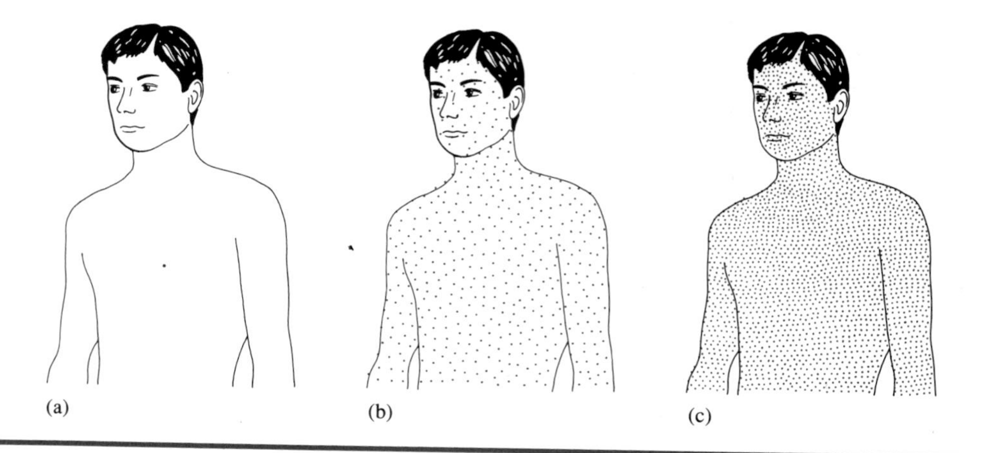

Figure 4.6 A comparison of relative amounts of lead in people. (a) Natural amount found in prehistoric Peruvians (one dot). (b) Average amount found in present-day Americans (500 dots). (c) An amount which will cause frank lead-poisoning in a significant fraction of a group of people (2000 dots). Each dot represents a unit of lead equivalent to 0.1 µg per gram of dry bone. The picture gives the impression that a modern human is closer to a victim of lead poisoning than to his/her prehistoric ancestor.

shows one that was used by Patterson himself. Its message is that, in terms of lead content, a modern human has much more in common with a victim of lead poisoning than with his or her prehistoric ancestor.

Activity 4.1

Clearly Figure 4.6 can be used to support the above argument. It does so by using dot density, which effectively draws attention to the *ratios* of the lead contents of the three people. But suppose that you wished to claim that person (b) is closer to a prehistoric human than to a victim of lead poisoning. How would you compare the three people then? Also, from what you have read so far in this chapter, is the level of bone lead in person (b) typical? If not, does this make much difference to the arguments? Finally, use your conclusions to construct a bar chart which gives a very different impression from Figure 4.6, and opposes the environmentalist case.

It is important to note that Patterson's and Nriagu's work did not go unchallenged. For example, Zbigniew Jaworowski of the Central Laboratory for Radiological Protection in Warsaw repeatedly published results in reputable scientific journals which were at odds with Patterson's. Jaworowski also concluded that anthropogenic emissions of lead are only 7% of the total, rather than the 97% claimed by Nriagu. Many of these arguments revolved around Patterson's very low figures for the levels of lead in prehistoric bones and ancient ices. Patterson claimed that the failure of others to confirm these was due to the contamination of their samples by pervasive lead pollution in the considerable period between sample collection and analysis, and to less sensitive analytical techniques. He himself took elaborate precautions to avoid contamination, lining his laboratory with plastic, and fitting it with an air-lock. Patterson also had a very high reputation in the field of isotope dilution analysis which, at the level of accuracy that he aimed for, is practiced in very few laboratories. This reputation was partly founded in the 1950s when he had used isotope dilution analysis of lead isotopes in rocks to obtain the accepted figure for the age of the Earth (4 550 000 000 years). When scientific findings disagree, it is often only those who have the time, interest and expertise to repeat the work who can decide between them. The number of such people is usually small, and may even be non-existent. In these circumstances, authority and reputation count for a great deal, and Patterson had them both. The Royal Commission included his work and Nriagu's in their report, but did not mention Jaworowski's:

> *...man's activities have resulted in increased lead concentrations in nearly all environmental media, particularly the air. The extent of this increase is greater than previously believed, for recent work, using ultra-clean techniques, has demonstrated that most of the estimates of natural lead concentration in the environment made before the last decade were too high because of contamination during sampling and analysis. Many studies have shown how lead levels have risen over the whole of the Earth's surface, and how 20th century man is having a profound effect on the amount of lead in his environment..., we are convinced that it would be prudent to reduce further its anthropogenic dispersal, and man's exposure to it, and we so* ***recommend****.* (paras 2.2 and 5.25)

As you will see in Chapter 7, the Royal Commission's deference towards Patterson's results seems to have been justified by subsequent events.

Question 4.1 A 30 kg sample of Arctic ice contains trace quantities of lead which has a normal isotopic distribution, i.e. 52% ^{208}Pb, 21% ^{207}Pb and a $^{208}Pb/^{207}Pb$ ratio of 2.5. The ice is melted, and 40 μg of pure ^{208}Pb is added in the form of $Pb^{2+}(aq)$.

After several hours, the lead in the solution is precipitated as lead sulphide which is then subjected to isotopic analysis in a mass spectrometer. The peak heights for ^{208}Pb and ^{207}Pb are 130 and 2 respectively.

(a) Calculate the original concentration of lead in the ice.

(b) If this were a sample of the type used in putting together Figure 4.5, when would you expect the ice to have been deposited?

4.3 Lead, child behaviour and child intelligence

In Section 2.4.1 you saw that symptoms of frank lead poisoning have occasionally been observed when the victim's blood lead concentration is only about $70\,\mu g\,dl^{-1}$. Few people dispute the fact that lead levels in this range are grounds for medical treatment. But at the cutting edge of their campaign against leaded petrol, the environmentalists put the dangers of lead levels well below this range. They argued that at the then acceptable levels of lead in teeth and blood, the development of young children was being impaired.

Table 4.2 The numbers of children in the six groups of different tooth lead concentration used in the study by Needleman in 1979.

Group	Range in tooth lead/$\mu g\,g^{-1}$	Number in group
1	<5.1	146
2	5.1–8.1	378
3	8.2–11.8	549
4	11.9–17.1	549
5	17.2–27.0	378
6	>27.0	146

What was needed to push this case forward was a study of a large typical sample of young children which suggested that those with higher lead burdens were more likely to be less intelligent and more poorly behaved. Just such a study was published by Herbert L. Needleman and others in 1979, when both intelligence and behaviour were examined. In the work on behaviour, 3 329 six- and seven-year-old children at schools in two towns in Massachusetts, USA, were asked to hand over shed teeth to their teacher; 2 146 of them obliged, and the concentrations of lead in those teeth were determined. The children were then divided up into six groups according to the concentration of lead in their teeth. The ranges of lead concentration and the numbers for each group are shown in Table 4.2. As the figures suggest, the ranges were chosen so that equal numbers fell into groups 1 and 6, into groups 2 and 5 and into groups 3 and 4.

Each pupil's teacher was then asked to complete a questionnaire on the child's behaviour by answering either 'yes' or 'no' to the 11 questions in Table 4.3.

Table 4.3 The questions that teachers were asked in Needleman's study of child behaviour in 1979.

1	Is this child easily distracted during his/her work?
2	Can he/she persist with a task for a reasonable amount of time?
3	Can this child work independently and complete assigned tasks with minimal assistance?
4	Is his/her approach to tasks disorganized (constantly misplacing pencils, books etc)?
5	Do you consider this child hyperactive?
6	Is he/she over-excitable and impulsive?
7	Is he/she easily frustrated by difficulties?
8	Is he/she a daydreamer?
9	Can he/she follow simple directions?
10	Can he/she follow a sequence of directions?
11	In general, is this child functioning as well in the classroom as other children his/her own age?

▷ Which of the two answers to question 1 in Table 4.3 is a mark of undesirable behaviour?

▶ The answer 'yes'; the child is easily distracted.

▷ Which of the two answers is a mark of undesirable behaviour when answering question 2 in Table 4.3?

▶ The answer 'no'; this implies that the child does not persist with tasks that he/she has been asked to do.

Thus, depending on the question, undesirable behaviour may be marked by a 'yes' or a 'no' answer. These undesirable answers are called *negative ratings*. Figure 4.7 shows the distribution of negative ratings among the six groups of Table 4.2 for each of the 11 questions in Table 4.3.

The data in Figure 4.7 are presented in the form of histograms. The figure contains 11 histograms, or sets of six bars. The 11 sets represent the 11 questions of Table 4.3. The six bars in any set, marked 1–6 in Figure 4.7, correspond to the six groups of children of different tooth lead levels. These levels are tabulated as ranges in Table 4.2. The height of a bar, when measured against the scale on the vertical axis to the left, then gives the negative ratings as a percentage of the total number of answers for that group to the question at issue. Activity 4.2 asks you to interpret this figure, and is a little more complicated than your encounter with histograms in Activity 2.2.

Activity 4.2

(a) What percentage of children in the highest-lead group were judged by their teachers to be easily distractible?

(b) What percentage of children in the lowest-lead group were judged by their teachers to be easily distractible?

(c) What percentage of children with tooth lead concentrations in the range 8.2–11.8 $\mu g\, g^{-1}$ were judged by their teachers to be easily distractible?

(d) How many children in the highest-lead group, and how many with tooth lead in the range 8.2–11.8 $\mu g\, g^{-1}$ were judged by their teachers to be easily distractible?

(e) What percentage, and how many of the children in the group with tooth lead concentrations in the range 5.1–8.1 $\mu g\, g^{-1}$ were judged daydreamers by their teachers?

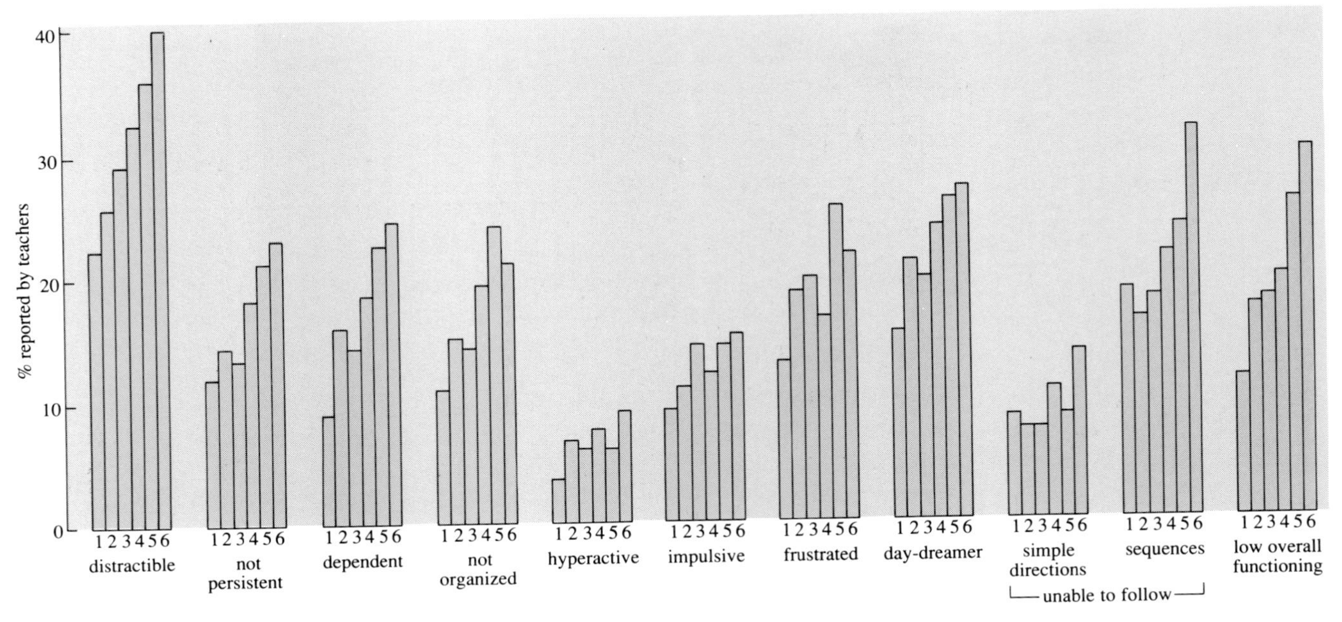

Figure 4.7 The percentage of negative ratings received by the six groups of children in Table 4.2 when their teachers answered the 11 questions in Table 4.3.

A mere glance at Figure 4.7 suggests that, for most of the questions in Table 4.3, there is a definite association or correlation between the percentage of negative ratings obtained in a group, and the group's tooth lead concentration. Strictly speaking, formal mathematical calculations are needed to confirm this, but here, the obvious tendency for the height of the bars to increase as one moves from group 1 to group 6 allows one to detect it by eye. Such a tendency strengthens the case for an association between lead levels and negative ratings because it is evidence of a **dose–response relationship**: of a *steady increase* in negative ratings with the lead dose.

4.3.1 Confounding variables

All Figure 4.7 does is to reveal that in Needleman's sample, there was an *association* or *correlation* between lead levels and negative ratings. It does *not* show that higher lead levels *cause* more negative ratings, any more than it shows that the undesirable behaviour reflected in the negative ratings causes the higher lead levels. Indeed, there is another very important possibility: there might be some other factor which influences both. Such a factor is known as a **confounding variable**, and in this case, an important confounding variable is social class. It is well known that children from poor homes are more likely to be exposed to old peeling paint and to be less closely supervised by their parents. Their immediate surroundings contain more lead, and they are given more opportunities to ingest it. At the same time, their home background leaves them ill-prepared for school, and they are less likely to be the teacher's favourite. Social class may therefore influence both lead levels and negative ratings, and explain the association between them. Detecting possible confounding variables and establishing causality are two of the most difficult problems in this type of statistical survey. Activity 4.3 clarifies this by using the example of two familiar human characteristics.

Activity 4.3

A statistician chooses a very large sample which is typical of the population as a whole (men, women and children). He measures heights and inspects heads, and finds that there is a significant association between height and baldness. Identify two confounding variables for which no allowance has been made.

Next, only the data on subjects aged 21 or more are retained. The association between height and baldness remains but is now much weaker. Why is this?

Finally, the data are restricted to males aged 21 or more. Now the association between tallness and baldness disappears. Indeed, there is a weak association between shortness and baldness. Why is this?

This example shows how the elimination of the effects of confounding variables may not just eliminate an association between two factors; it may even reverse it. So if social class influences both the negative ratings and the lead levels in Figure 4.7, attempts should be made to subtract that influence. This is a difficult thing to do. Differentiating people by social class is much harder than differentiating them by age or sex—the confounding variables used in our tallness/baldness example. As a researcher in this field has said, 'How much sense does it make to ask whether the difference between a judge and a school teacher is the same as the difference between a milkman and a labourer?'

Despite these problems, attempts have been made to subtract the influence of social class and other confounding variables on the association between lead levels and classroom performance or behaviour. Needleman did not attempt to do it for the six-group comparison in Figure 4.7. What he did instead was to compare, from among the sample of children, 100 with low lead levels ($<6\,\mu g\,g^{-1}$) and 58 with high lead levels ($>24\,\mu g\,g^{-1}$). The comparison was made by first subjecting these two groups to standard intelligence and other performance tests, and then trying to make allowances for the effects of confounding variables such as parental education and social class. This part of the work also had the merit of avoiding any reliance of the subjective response of the child's teacher. Even after allowing for social background, in almost every case the high-lead group performed significantly less well than the low-lead group. However, with just a two-group classification, the chance of detecting a dose–response relationship was lost. Needleman's results were obviously important; but would other similar work corroborate or conflict with his conclusions?

4.3.2 British studies since 1979

Table 4.4 gives data on five important studies of the relationship between some measure of the lead burden in young children, and their performance in intelligence tests. Performance is measured in terms of a number called an intelligence quotient or IQ rating which is effectively a mark scored when attempting a standard set of tests. All these studies made corrections for one or more important confounding variables associated with social class, such as parental occupation, parental education or parental IQ. If, after correction, a significant association between increasing lead levels and decreasing IQ was found, this is marked by an 's.' (significant) in the last column. If this was not found, the initials 'n.s.' (not significant) appear. **Significance** here is a statistical term which is asserted at a particular level of confidence. In the studies of Table 4.4, significance was tested by assuming first that there is no association between higher lead level and lower attainment. When a study is conducted and reveals such an association, one then calculates the probability of the association occurring by chance. If this probability is less than one in 20, the association is said to be significant at the 95% confidence level. To give you some idea of what this test means, we can compare it with ten tosses of a coin. The probability of getting eight heads is slightly less than one in 20, so asserting that an association is significant at the 95% confidence level is rather like suggesting that a coin is biased if one throws it ten times and gets eight or more heads. Clearly, it might not be biased, but this particular test suggests otherwise.

Table 4.4 British studies of the association between lead level and child attainment. Those interested in the original papers will find the references in the following review: Needleman, H. L. and Gatsonis, C. A. (1990) *Journal of the American Medical Association*, **263**, p. 673.

Study	Date	Place	Number of subjects	Tissue analysed	Significance	Author's name*
1	1981	London	166	blood	s.	Yule
2	1983	London	402	teeth	n.s.	Smith
3	1984	Birmingham	187	blood	n.s.	Harvey
4	1986	London	160	blood	n.s.	Lansdown
5	1987	Edinburgh	501	blood	s.	Fulton

* Only the first listed author is given.

▷ Which of the studies in Table 4.4 found an association between high lead level and low attainment which was significant in the sense described above?

▶ Numbers 1 and 5.

One is naturally drawn to the Edinburgh survey of 1987 because it is the most recent, and studied the largest sample. However, its results conflict with the London study 2 for which the sample size is also large. In 1987, the authors of study 2 used their data to calculate the predicted change in IQ when the lead level for a child was doubled. This quantity was chosen because it was also easily obtainable from the published data in the Edinburgh study. A comparison could therefore be made and Figure 4.8 shows the result.

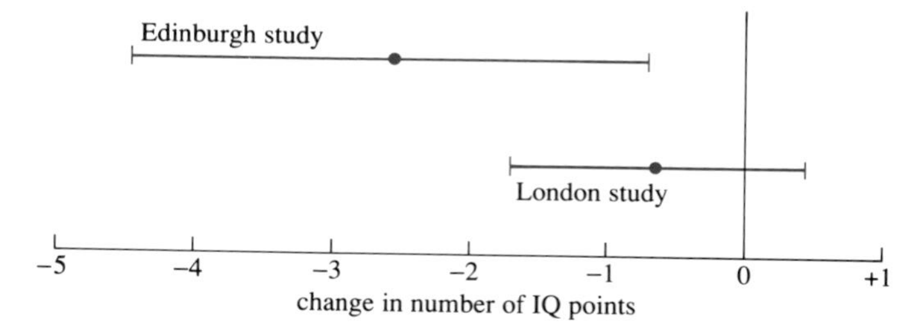

Figure 4.8 Estimated change in IQ for a doubling of body-lead level calculated from the Edinburgh study, 5, of Table 4.4, and from data in the London study 2. The dots indicate mean values for the change, and the horizontal lines mark the error limits calculated from the data of each survey. The error limits are 95% confidence intervals. They imply that if the surveys were repeated many times, then the result would fall between the limits on at least 95% of occasions.

▷ Is there any sense in which the studies are consistent?

▶ They are, in the sense that the two ranges of change in IQ which arise from uncertainties within the two sets of data, overlap in the −0.6 to −1.7 region.

In the words of the authors of the exercise, 'the two studies are consistent in pointing towards a weak inverse association* between body-lead burden and IQ in young children'. However, the authors also agreed that their procedure could be challenged because it combined two sets of results that are not strictly comparable.

▷ In what important respect, apart from location, are the two studies not comparable?

▶ The Edinburgh survey used blood lead as an index of exposure; the London survey used tooth lead.

As noted in Section 2.4, it is generally agreed that blood lead is a measure of recent exposure to lead, and tooth lead a measure of longer-term exposure. In this particular case, that difference leads on to a further reason for non-comparability of the Edinburgh and London studies. Although the Edinburgh work used blood lead as an index of lead exposure, some measurements of tooth lead were made as well. The Edinburgh childrens' teeth had unusually high lead concentrations, some $2\frac{1}{2}$ times those found in the London children. The difference has been attributed to an exposure to high water-lead levels in Edinburgh during infancy; as discussed in Section 2.6, water-lead has been a particular problem in Scotland where water tends to be softer. This further non-comparability of the two studies may mean that any conclusions drawn from the Edinburgh survey are less likely to be transferable to other parts of the UK.

All this shows how hard it is to come to any definite conclusions about the effect of lead on IQ. The authors of the Edinburgh survey claim only that their results 'add to the evidence that lead at low levels of exposure probably has a small harmful effect on the performance of children in ability and attainment tests'. This effect, they note, is much less than the effect of a factor like the parents' educational qualifications. They also point out that *association* between higher lead and lower IQ does not prove *causality*. There could be some unconsidered confounding variable which causes both high lead and lower IQ, or there could be *reverse causality*: a lower IQ could cause behaviour which leads to a greater lead intake. They remark that 'it is difficult to conceive of any

* Inverse association means that as one quantity (body-lead level) goes up, the other (IQ) goes down.

observational study that would enable us to untangle the various causal pathways in the lead/ability relation'. However, the authors of the London study 2 which found no significant relationship between lead burden and IQ later remarked:

> *Meanwhile, the evidence on lead–IQ associations cannot be lightly dismissed and it would seem prudent to continue active policies of reducing children's exposure to lead from all identified sources including water, petrol, food and paint.*

This was also the view of the Medical Research Council when it reviewed the research on the subject in the period 1984–1988. The evidence for harm is a little stronger than in 1983 when the Royal Commission report said:

> *In our view, the accumulated evidence may indicate a causal association between the body burden of lead and psychometric indices*, or the effect of confounding factors, or both. On present evidence we do not consider it possible to distinguish between these possibilities. We consider it unlikely that population surveys alone will settle the issue in the near future, particularly as the effect of lead on behaviour and intelligence, as suggested by the studies mentioned above, is at the most small at the concentrations found in the general population.* (para. 5.12)

Question 4.2 In Sections 4.3.1 and 4.3.2, you saw that if an association exists between high lead burden and low attainment, causality does not follow: one alternative possibility is reverse causality—lower attainment may be a cause of higher lead uptake. Does this ambiguity also hold in the association between maleness and baldness that emerged from Activity 4.3?

4.4 The contribution of leaded petrol to a person's lead uptake

Just what is the contribution of leaded petrol to the lead that we take in from our environment? One well-known attempt to answer this question was the so-called Turin experiment. In Section 4.2, we spoke of the 'normal' abundances of lead isotopes; these are about 52% ^{208}Pb, 21% ^{207}Pb, 25% ^{206}Pb, and 1–2% ^{204}Pb. This corresponds to a $^{206}Pb/^{207}Pb$ ratio in the range 1.16–1.20. However, some localized samples of lead ore have ratios rather different from this value. For example, lead from Broken Hill in Australia has a $^{206}Pb/^{207}Pb$ ratio of 1.04.

In the 1970s, Italian lead alkyls were manufactured by just one company. Thus between 1975 and 1979, it was possible to arrange for all Italian petrol to contain lead alkyls made only from lead from Broken Hill. A study was then made of the changes in isotopic composition of airborne lead, and of the blood lead of certain individuals, in the area around Turin in northern Italy (Figure 4.9).

Activity 4.4

By using Figure 4.9, compare the isotopic ratio for leaded petrol in 1974 with the average measured values in the period mid-1977 to mid-1979. Then compare the change with the corresponding change for airborne lead. What does the comparison tell you about the contribution of leaded petrol to total airborne lead?

In Figure 4.9, the decrease in the ratio is much smaller for blood lead than for airborne lead. The ratio falls from just over 1.16 to an average of about 1.14 between

* A psychometric index is some measure of mental ability.

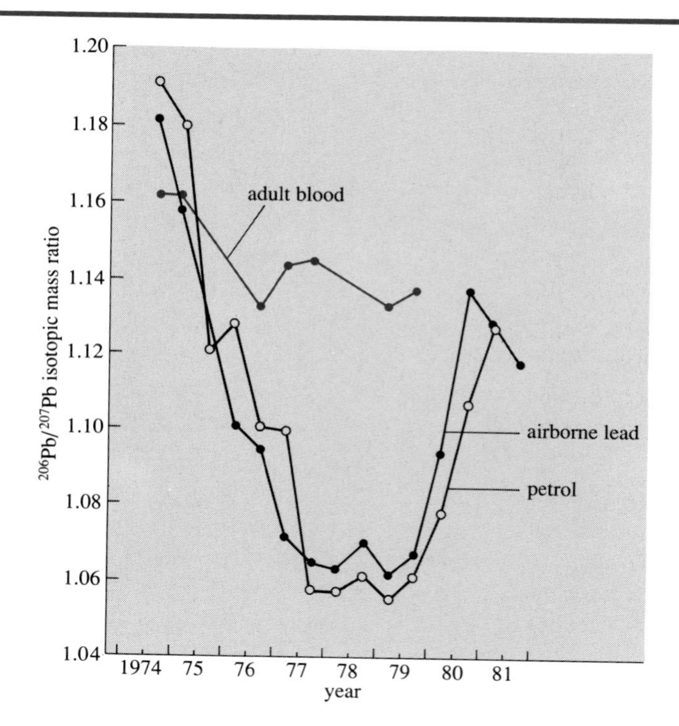

Figure 4.9 Changes in the mean $^{206}Pb/^{207}Pb$ ratio in petrol, airborne lead and adult blood between mid-1974 and mid-1981 in Turin. Leaded petrol with the ratio 1.04 was progressively introduced in the period 1975–9.

1976 and 1979. This decrease is about 20–25% of the decrease to 1.06 that would have been expected if all blood lead were derived from leaded petrol. Thus in central Turin at this time, petrol seems to have been responsible for about 20–25% of blood lead. In the surrounding rural areas, the value was smaller—about 12%.

The Turin experiment has been criticized because the sample was small and not properly representative. Nevertheless, its conclusions were not very different from other studies of the same phenomenon, and the Royal Commission did not take issue with them:

> *The contribution that petrol makes to the uptake of lead has been a matter of particular concern... On present evidence, for large numbers of adults in the UK it contributes at least 20% of their uptake; for the majority it contributes less.*
>
> (paras 4.51 and 8.25)

In the case of children, the report was reluctant to state a figure because of uncertainties about how much lead from leaded petrol a child took up through the ingestion of dust.

The Commission's 20% figure included an estimate of the contribution made through the deposition of petrol lead on soil, and its subsequent uptake by plants. It is worth reiterating the obvious point that if, in the late 1970s, 20% of the city dwellers' lead uptake came from leaded petrol, 80% of the uptake did not. As discussed in Sections 2.4 and Figure 2.6, this 80% must come primarily from ingested food and water.

4.5 Vehicle exhaust emissions and the catalytic converter

The British debate about leaded petrol was conducted partly in the shadow of legislation about vehicle exhaust gases that had already been passed in the USA. In Section 3.1, the combustion of octane was the model for the chemical reactions that take place in a petrol engine:

$$2C_8H_{18}(l) + 25O_2(g) \longrightarrow 16CO_2(g) + 18H_2O(g) \qquad (3.1)^*$$

The air–fuel mixture is usually controlled so that there is enough oxygen to convert all the hydrocarbon fuel into carbon dioxide and water vapour as in Equation 3.1.

Unfortunately, however, the mixing of the hydrocarbons and oxygen in the cylinder is never so perfect that this conversion is completed in the brief span of the ignition stroke. Some of the hydrocarbons fail to react with the oxygen or react with it only partially, and the oxygen that is not consumed in this way may do other things instead. The result is that, although the exhaust gases (ignoring nitrogen and oxygen) consist mainly of carbon dioxide and water vapour, other things are present as well. Three of these ingredients are especially important:

1 Unburnt hydrocarbon vapours which have failed to react with oxygen.

2 Carbon monoxide, CO; in the short span of the ignition stroke, the carbon atoms in some of the hydrocarbon molecules gain neither the time, nor sufficient access to oxygen, to be converted to carbon dioxide, CO_2. They only take on half the necessary oxygen and form carbon monoxide instead.

3 Oxides of nitrogen such as NO and NO_2; the temperature inside the cylinder during the ignition stroke may reach 2200 °C. At this temperature, nitrogen gas, N_2, in the air which is drawn in with the fuel may combine with oxygen to form nitric oxide, NO:

$$N_2(g) + O_2(g) \longrightarrow 2NO(g) \qquad (4.2)$$

When nitric oxide leaves the exhaust system and cools down, it reacts with atmospheric oxygen to form the brown poisonous gas, nitrogen dioxide, NO_2:

$$2NO(g) + O_2(g) \longrightarrow 2NO_2(g) \qquad (4.3)$$

4.5.1 Some harmful effects of exhaust emissions

None of the above ingredients is a desirable addition to the atmosphere. As you saw in Section 3.4, some of the unburnt hydrocarbons, especially aromatics such as benzene and PAHs, can cause cancer. Carbon monoxide is a very poisonous gas. Oxides of nitrogen, too, are toxic, and they are also an important source of acid rain. But the three emissions are most notorious in a special combination of climate and geography. Sheltered by surrounding hills, the city of Los Angeles enjoys more windless hours of sunshine than most places. Its still atmosphere is often infiltrated from below by the exhaust gases of millions of motor vehicles, and simultaneously irradiated by sunlight from the cloudless skies above. In the 1960s, the sunlight set in motion chemical reactions between oxides of nitrogen and unburnt hydrocarbons. The result was a persistent chemical fog, or smog, in which breathing sometimes became uncomfortable.

To stop this happening, Californians decided to administer the *coup de grace* to all three pollutants by destroying them chemically in the vehicle's exhaust system. This policy was adopted first by the state of California in the 1960s, and was later embodied in the US Clean Air Act of 1970 (see below).

4.5.2 The three-way catalytic converter

If the exhaust gases from a petrol engine contain the right amount of atmospheric oxygen, then the ingredients of chemical reactions that can destroy the three important classes of pollutant are present. For example, the combustion of the hydrocarbons which was meant to occur in the engine might still take place:

$$2C_8H_{18}(l) + 25O_2(g) \longrightarrow 16CO_2(g) + 18H_2O(g) \qquad (3.1)^*$$

The carbon monoxide can be destroyed by oxygen:

$$2CO(g) + O_2(g) \longrightarrow 2CO_2(g) \qquad (4.4)$$

Alternatively, it and the oxides of nitrogen can suffer mutual annihilation, forming carbon dioxide, and harmless nitrogen gas:

$$2NO(g) + 2CO(g) \longrightarrow 2CO_2(g) + N_2(g) \qquad (4.5)$$

It is known that the equilibrium in the three systems represented by Equations 3.1, 4.4 and 4.5 lies to the right-hand side—it favours the products of the reaction, and therefore the destruction of the pollutants. The drawback is that in the exhaust system, the reactions occur too slowly, and the pollutants are swept out into the atmosphere before the reactions take place. As noted in Section 3.3.2, this is just the sort of situation which can be put right if we can find the right catalysts.

The catalysts currently used to speed up reactions shown in Equations 3.1, 4.4 and 4.5 are the metals platinum and rhodium. They are among the most expensive of the class of chemical elements known as transition elements. Platinum speeds up the oxidation reactions shown in Equations 3.1 and 4.4 which destroy the hydrocarbons and carbon monoxide.

▷ So what, by a process of elimination, does the rhodium destroy?

▶ Oxides of nitrogen; it does this by promoting reactions such as that of Equation 4.5.

Figure 4.10 A three-way catalytic converter. The metal shell has been partially cut away, exposing a gauze lining, inside which is the cylindrical grid of exhaust channels. A separate grid of this type is shown above and to the left. It is black because the platinum–rhodium catalyst has been dispersed over its surfaces. Before the catalyst is spread over it, the ceramic grid is white, as shown above and to the right.

One device in which the clean-up process can occur is called a **three-way catalytic converter**. This is inserted into the exhaust system between the engine and the silencer (Figures 4.10 and 4.11) and is called 'three-way' because it destroys all three classes of pollutant. For each class, destruction levels of 90% or more are achievable.

From the standpoint of this book, the crucial point about three-way and other catalytic converters is that their efficiency is greatly reduced by the lead compounds in the exhaust stream when a car engine burns leaded petrol (Section 3.3.4). For example, solid lead halides, such as $PbBr_2$, and lead sulphate, $PbSO_4$, can block the pores on the aluminium oxide granules where the metal particles sit. The lead, in fact, is said to *poison* the catalyst. Thus catalytic converters are seriously damaged by leaded petrol, and both unleaded petrol, and cars able to use it, are necessary if exhaust systems are to include converters of this type. In the USA, concern over phenomena like Los Angeles smog led, in 1970, to a Clean Air Act which called for cars introduced in 1975 and subsequent years to be fitted with such converters. The act stipulated, at the same time, that all major gasoline outlets would have to sell at least one grade of unleaded petrol to provide for these new models. Figure 4.12 shows the effect that this Act, and subsequent modifications have had on the consumption of lead in petrol in the USA in recent times.

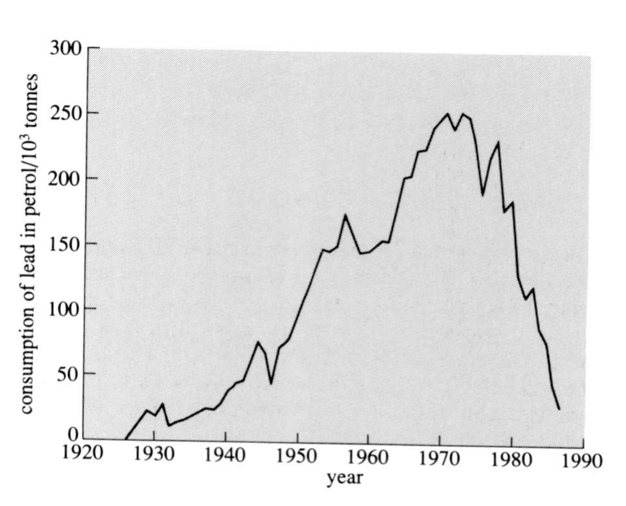

Figure 4.12 The consumption of lead in petrol in the USA during this century.

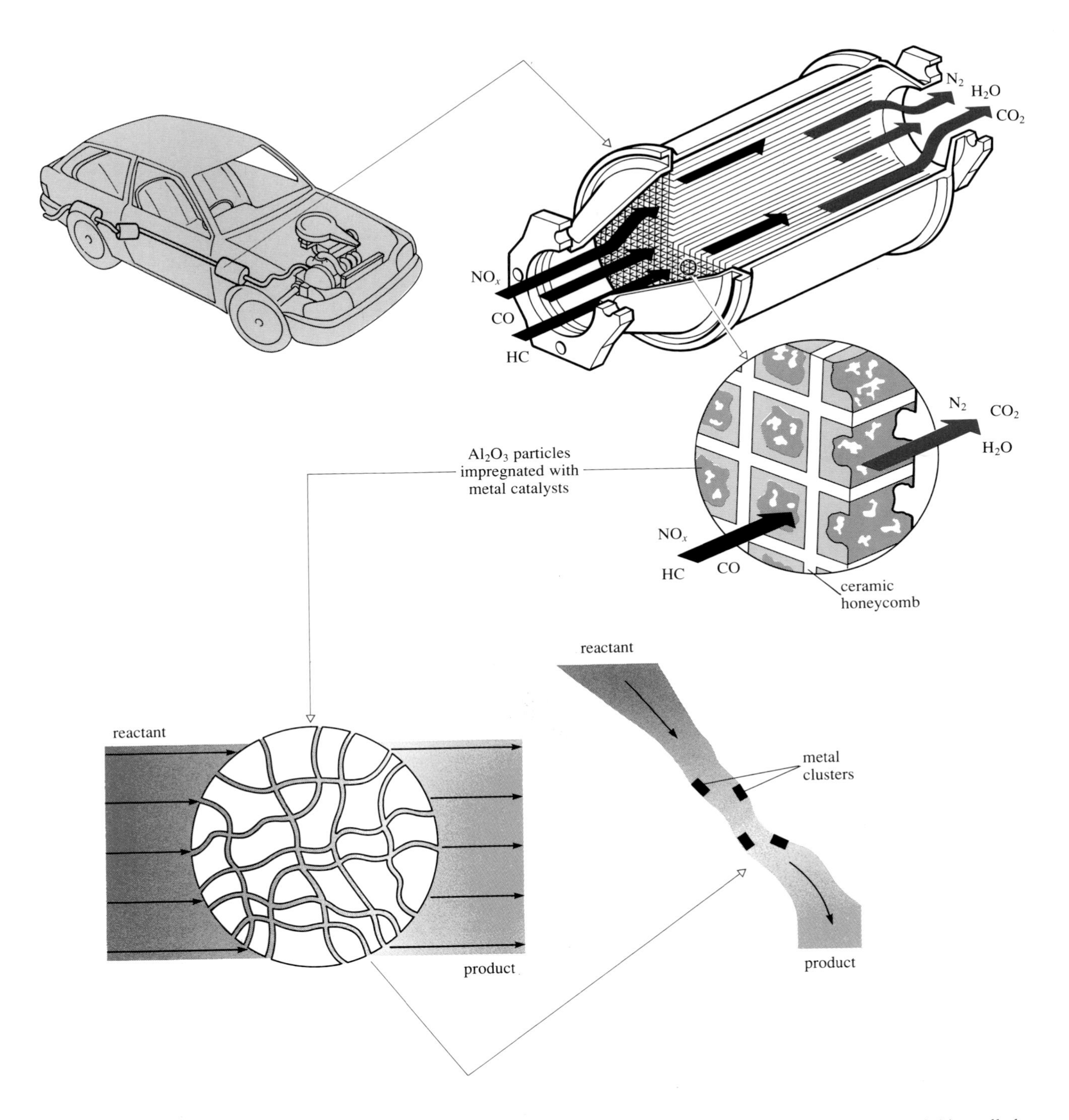

Figure 4.11 The structure of one type of three-way catalytic converter. The core consists of a cylindrical grid of thin-walled channels of square cross-section, composed of a ceramic material made from oxides of magnesium, aluminium and silicon. The platinum–rhodium catalyst is dispersed over granules of solid aluminium oxide, Al_2O_3, which have been specially prepared with a high surface area. The catalyst-coated granules are mixed with water to form a slurry, and passed through the grid, which is then heated in a furnace. The process leaves Al_2O_3, impregnated with catalyst particles, dispersed on the walls of the channels. In passing through the channels, exhaust pollutants traverse pores in the Al_2O_3 granules, encountering metal catalyst sites where reactions such as those shown in Equations 3.1, 4.4 and 4.5 occur. Efficient conversion occurs only if the air–fuel ratio on entry to the converter is right. The ratio is controlled by measuring the oxygen with a sensor and then making any necessary adjustments to the air and fuel supply. (NO_x denotes oxides of nitrogen; HC denotes hydrocarbons.)

When the Royal Commission were preparing their report, they were conscious that similar legislation was under consideration by an EC working group. If exhaust catalysts were recommended, then leaded petrol would begin to look like an anachronism. One of the Commission's final conclusions was:

> *The development of engines to run on unleaded petrol would remove an impediment to the control of other gaseous emissions by the fitting of exhaust catalysts, a requirement currently being considered by the Commission of the European Communities, already demanded by the USA and Japan, and shortly to be demanded by Australia.* (para. 8.38)

This was good anticipation. Currently (1992), the projected EC exhaust emission standards for carbon monoxide, hydrocarbons and nitrogen oxides are such that exhaust catalysts will be needed to meet them: by the end of 1992, exhaust catalysts should be fitted to all new EC cars.

Question 4.3 The driver of a car fitted with a three-way catalytic converter switches from the recommended unleaded petrol to a leaded brand. As a result of this change, do the *amounts* of the following exhaust gases increase or decrease: (a) carbon monoxide, (b) carbon dioxide, (c) nitric oxide, (d) hydrocarbons, (e) nitrogen.

Summary of Chapter 4

1 In the 1970s, the average blood lead concentration in the population was about one-quarter of that at which symptoms of frank lead poisoning have sometimes been observed.

2 Isotope dilution mass spectrometry has been used to reveal increases in the lead content of Arctic snow since 800 BC, and in the lead content of bones since prehistoric times. The results are widely held to show that as a result of lead technologies, such as plumbing, smelting and the use of leaded petrol, the modern world exposes people to very much greater concentrations of lead than did a primitive prehistoric existence.

3 There is evidence that lead levels well below those at which symptoms of frank lead poisoning occur, may have small, adverse effects on child attainment and behaviour. This evidence, however, is not unequivocal, mainly because of uncertainties caused by the possible influence of confounding variables, one of which is social class.

4 Changes in isotope ratios when leaded petrol with an unusually low $^{206}Pb/^{207}Pb$ ratio was introduced in Italy suggested that, in Turin, petrol was responsible for nearly all airborne lead, and for about 25% of adult blood lead.

5 Metallic platinum and rhodium are used as catalysts in three-way catalytic converters to destroy hydrocarbons, carbon monoxide and nitrogen oxides. They are poisoned by lead compounds in leaded petrol exhaust streams.

5 Generating concern about leaded petrol: the political campaign

This chapter describes how the scientific arguments of Chapter 4 were woven into political activities during the campaign against leaded petrol. It begins, in a small way, to develop skills of abstracting and organizing information from more than one source that will become more important in later books of the Course. When you have read the chapter you will be asked to compare its arguments and emphases, first with those of Chapter 4, and then with the ones used in a short paper from a scientific journal. These exercises will provide your first sustained encounter with higher-level skills such as 8 and 9 on p. 77, that will become increasingly important in later parts of the Course.

Chapter 5 does not pretend to be a full account of the political exchanges that took place. Rather, its purpose is to raise questions that will recur throughout *Science Matters*, and which you must ultimately answer for yourself. For example, the scientific attitude is often said to consist of a determination to judge explanations *solely* by how well they account for the results of experiments; is such an attitude compatible with political activity in which science plays a part? Again, how is science altered when it moves from the arena of scientific papers and conferences into newspapers and on to television? In this chapter, extracts from newspapers will provide you with the chance to think about this last question in particular. However, they are mainly illustrative of the main text, which can be read independently of them, and indeed, it is only the main text that you need be acquainted with when you complete Activity 5.1 at the end of the chapter.

Extract 5.1 From *The Times*, 15 February 1971.

Danger from lead in the air, says professor

By a Staff Reporter

The amount of lead in the blood of people living in industrial countries could be at dangerous levels, Professor D. Bryce-Smith, Professor of Organic Chemistry at Reading University, says in an article in *Chemistry in Britain*.

Warning the public of the dangers from airborne lead emitted by the exhausts of vehicles he says he knows of no other toxic pollutant that has accumulated in man 'to average levels so close to the threshold for overt clinical poisoning'.

He told *The Times* last night: 'Children in particular are at risk because the resultant brain damage to them tends to be permanent, whereas adults can recover. The lead levels in petrol have been tending to rise in the past 10 years, and the total amount getting into the atmosphere is increasing'.

Airborne lead was more dangerous than lead taken in the diet because of the amount inhaled. 50 per cent was absorbed into the body fluids, whereas of the amount taken in diet, only 5 to 10 per cent went into the body fluids.

People who lived in cities should not grow their own vegetables because there was too much lead in the air.

5.1 The beginnings

In Britain, leaded petrol first became a political issue when Professor Derek Bryce-Smith of Reading University published two papers in the journal, *Chemistry in Britain* in 1971. The first of these is reproduced in Chapter 6, and later on you will be asked to read it. Bryce-Smith's claims quickly gained him access to the newspapers (Extract 5.1), and the issue received wide coverage in more popular scientific journals such as *New Scientist*. It was also debated on the BBC TV programme, *Controversy*, in August 1971. Extract 5.2 is a review of this programme. Opponents of Bryce-Smith and of those who supported him, found themselves at a disadvantage because, as one *New Scientist* commentator wrote, 'public opinion is swinging towards the idea that pollutants are guilty until proven innocent'. Consequently, pressure began to build up on the Government to take action.

In 1972, the maximum permitted lead content of petrol was $0.84\,g\,l^{-1}$. In that year, the Government announced a phased programme of reductions beginning with a fall to $0.64\,g\,l^{-1}$ in 1973 (Extract 5.3). This was politically attractive because the existing figure of $0.84\,g\,l^{-1}$ was unnecessarily high: the highest lead level in 1972 petrol was already below $0.6\,g\,l^{-1}$, so the 'reduction' to a new legal limit of $0.64\,g\,l^{-1}$ was, in practice, business as usual. Subsequent reductions were put back relative to the

Extract 5.2 From *The Times*, 31 August 1971.

Controversy

BBC2

Chris Dunkley

Last night's programme in the BBC's *Controversy* series was, in a sense, quite extraordinarily irresponsible. The producers offered us a discussion, which in 50 minutes, attempted to convey to the layman a complete outline of the growing debate on lead pollution but naturally succeeded in doing no such thing: instead, it managed to suggest some very scary effects which may possibly be caused by lead in our diets and in the atmosphere, and left us with no very clear idea as to whether we should believe them.

I can hear them all down at the BBC Club now muttering 'if we knew the answers for certain there wouldn't be a debate, would there, and we wouldn't make a programme would we'. True, but no answer to the charge of irresponsibility. There was only one way this programme could have been turned into a responsible affair, and that would have been to allow it enough time for the protagonists to deal with the subject to their own satisfaction within some sort of responsible limit set by the chairman (in order to avoid it running for a fortnight). In other words it should have been open ended and, if necessary, run for three or four hours.

After all, if most of us are either to die from lead poisoning or suffer illness from it in varying degrees as was suggested in *Controversy* it would not seem unreasonable that the country's foremost news and education medium should spend one eighth of a day in a year discussing it. The BBC, however is just not that brave.

But the ridiculous sprint at which we were carried over the subject was not the programme's only drawback; it also offered a salutary lesson in the extremes to which the increasingly paranoid screeching for 'balance' is pushing the BBC. Having given Professor Derek Bryce-Smith his head, and allowed him to precis his views on the dangers of lead uninterrupted, it was apparently felt that 'balance' could only be sustained by lining up against him no less than six opponents, most of them (and possibly all—it was never made clear) with a vested interest in lead. Seven participants and 50 minutes to spare—how could one hope for any sort of coherent outcome.

What we actually saw was a repetition of the usual pattern of scientific debate on television: a man of passionate convictions on one side, trying desperately to involve the public in his beliefs and anxieties, and on the other a battery of hostile witnesses seeking not to broaden the debate, or find the truth, but intent simply on picking minute holes in an Aunt Sally.

Extract 5.3 From *The Times*, 2 August 1972.

Lead content to be halved

By Roger Vielvoye

The maximum lead content of British petrol will be reduced by nearly half during the next $3\frac{1}{2}$ years, Mr Peter Walker, Secretary of State for the Environment, announced yesterday.

The maximum permitted level of 0.84 grams of lead a litre will be reduced to 0.64 grams by the end of this year; to 0.55 grams by the end of 1973 and 0.45 grams by the end of 1975.

schedule in Extract 5.3 because of the 1973 oil crisis, but they were more effective, and provided for a maximum lead content of $0.4\,g\,l^{-1}$ by January 1981 (Figure 5.1). The total *real* effect of this programme was therefore to reduce lead content by only about 30%. Nevertheless it succeeded in sapping the agitation over leaded petrol for several years.

The agitation was powerfully renewed in 1978, when news leaked out of a study performed by a research group at the Atomic Energy Research Establishment at Harwell (Extract 5.4). The results suggested that the contribution made by airborne lead to the concentration of lead in blood was some 2–3 times greater than the group had previously believed. At the same time, news of Needleman's work (Section 4.3) began to trickle in from across the Atlantic, and Bryce-Smith and others set up an organiza-

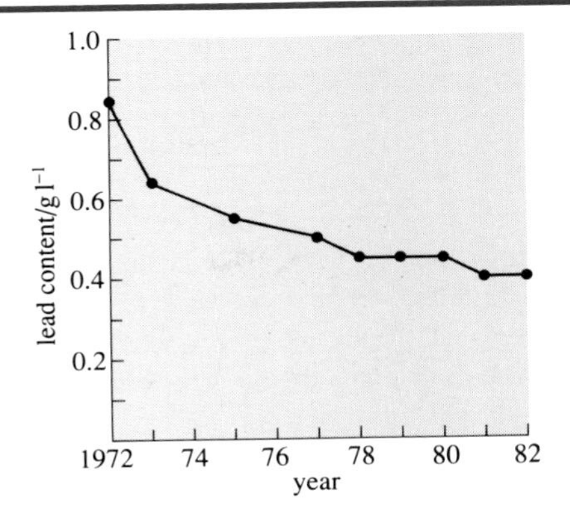

Figure 5.1 Changes in the maximum permitted lead content of British leaded petrol between 1972 and 1982.

tion called CALIP (Campaign Against Lead In Petrol). The Government's response was to set up two committees. The first was given the acronym WOPLIP (WOrking Party on Lead In Petrol) and its brief was to examine the costs and implications of changes to the legislation on leaded petrol. The second was a committee chaired by Patrick Lawther, the Professor of Environmental and Preventative Medicine at St. Bartholomew's Hospital, London. It was asked to look into the possible connection between environmental lead and health. By late 1979, pre-publication reports of both committees were in the Government's hands.

Extract 5.4 From *The Times*, 29 November 1978.

Lead poisoning report is 'leaked' at road inquiry

From Our Correspondent
Guildford

A government report which shows that absorption of lead from exhaust gas is far higher than ever realized before was 'leaked' at a motorway inquiry yesterday.

Dr Robert Stephens, reader in organic chemistry at Birmingham University produced the report by the Harwell Atomic Engineering Research Establishment at the M25 Leatherhead interchange inquiry.

Dr Stephens, aged 49, told the inspector, Rear-Admiral Harry Nixon, that the report confirmed his own research into lead poisoning. The Government had always maintained that only 10 per cent of all lead poisoning came from the air, but the Harwell report showed a completely different picture, he said.

He quoted the report as saying that for people living near a motorway 'the contribution of lead to total uptake must be comparable with that for diet'.

He continued, 'In other words, they have now conceded that half the intake of lead is from exhaust fumes. In fact I believe that, given one or two more years of research, it will be found that the pollution from the air is far greater than that'.

5.2 The WOPLIP report

When the WOPLIP report reached the Government in 1979, the lead content of 4-star, 97 RON petrol was close to $0.4\,g\,l^{-1}$ which by 1981 was due to become an upper limit. WOPLIP presented five options, but only the three most relevant are considered here. They were as follows:

1 Maintain a lead content of $0.4\,g\,l^{-1}$ and fit lead filters to the exhaust systems of new cars.

2 Cut lead content to $0.15\,g\,l^{-1}$ and maintain an octane number of 97 RON.

3 Market 92 RON unleaded petrol and require new cars to be designed to run on it. Existing cars can continue to use 97 RON petrol with a lead content of $0.4\,g\,l^{-1}$.

Extract 5.5 From *The Times*, 17 December 1974.

Violent teenage behaviour, including football hooliganism, might be attributed partly to lead pollution, Professor D. Bryce-Smith, Professor of Organic Chemistry at Reading University, said yesterday.
In a joint article in *The Ecologist* magazine with Professor Tony Waldron of the Department of Social Medicine at Birmingham University, Professor Bryce-Smith says that children with a condition known as hyperactivity have a tendency to violent destructive action. They quote evidence that hyperactivity may be linked to raised blood-lead levels.

Extract 5.6 From *The Times*, 29 November 1978.

Gulliver Handley's parents, and those of Fidel Budden and Merlyn Albery-Speyer, both aged two, are attempting to sue Shell, BP, the Ford Motor Company and Associated Octel, which makes lead additives for petrol, alleging that their children's health is being damaged by lead in petrol fumes.
They claim £1 damages for trespass, negligence and nuisance, and a further £1 special-damages for the inconvenience of not being safe near main roads.
Mr Pedley said a suggestion that the plaintiffs had embarked on the action only to obtain publicity was wrong. They were anxious to gain judgment on a matter of vital public interest.
The review continues tomorrow.

The cost of each option was expressed as an annual cost, at 1978 prices, during a future year some 10–20 years after the option has been introduced. For option 1, this cost was £72 million, for option 2, it was £124 million, and for option 3, it was £199 million.

Although option 1 was the cheapest, there were good reasons for avoiding it. Over their full lifetime, lead filters were likely to be only about 50% efficient. Moreover, when the car was scrapped, the lead might well leak back into the environment, unless the filters were carefully disposed of. There was also something faintly ridiculous about putting something that was not absolutely essential into one end of a vehicle, and installing a device at the other end to take it out again. Option 2 on the other hand was much more attractive. It was less expensive than option 3, and its costs would fall upon the oil industry rather than upon the car industry. In 1979, this was attractive, because British Leyland and other sectors of the UK motor industry were in a vulnerable state. Moreover, because option 2 was less drastic than option 3, it could be introduced more easily, and presented as the most effective way of achieving a quick reduction in airborne lead. Much depended however on the attitude struck by the Lawther Committee which was to look into several years of claims and counter-claims, some of which were quite riveting (Extracts 5.5 and 5.6).

5.3 The Lawther report

The Lawther Committee's brief was 'to review the overall effects on health of environmental lead from all sources and, in particular, its effects on the health and development of children and to assess the contribution lead in petrol makes to the body burden'. It devoted considerable space to Needleman's 1979 study and related work, pointing out that corrections for confounding variables were made after a division into just two groups, when a division into three might have elicited a dose–response relationship. The report also drew attention to an omission in Needleman's corrections for confounding variables and pointed to the possibility of reverse causality. The committee felt unable to come to clear conclusions, remarking:

> *Together these studies... provide some evidence of an association between raised tooth dentine lead levels and a slight lowering of measured intelligence. There are a number of reservations about these studies and the inferences to be drawn from them which in our view weakens their conclusions.*

In its discussion of blood lead levels, the committee noted that 'no firm evidence exists for guidance on what might be taken as a wholly *safe* level in the environment of the body'. It fell back on the upper figure specified in the 1977 EEC Directive on exposed populations. The Directive calls for remedial action if more than 2% of the population have blood lead concentrations above $35\,\mu g\,dl^{-1}$. The committee pointed out that 'there is no convincing evidence of deleterious effects at blood lead concentrations below about $35\,\mu g\,dl^{-1}$'. On the other hand, 'symptoms of lead poisoning and encephalopathy occur with levels in excess of, say $80\,\mu g\,dl^{-1}$... It is therefore in the range of blood lead concentrations between 35–$80\,\mu g\,dl^{-1}$ that doubt remains'. The committee therefore recommended that blood lead levels over $35\,\mu g\,dl^{-1}$ should be a trigger for action on the victim's behalf.

The report concluded that a typical contribution of inhaled airborne lead from petrol to blood lead levels was about 10%. For most people, food was the major source of lead, and the committee found 'no evidence that this is substantially enhanced by airborne lead'. With regard to action on leaded petrol, the committee recommended that the annual mean concentration of lead in air should be kept below $2\,\mu g\,m^{-3}$ in places where people were liable to be continuously exposed. It called for 'emissions of lead to the air from traffic and other sources to be progressively reduced, subject to an appraisal of any other effects on health of altering the constituents of petrol'.

To the surprise of at least some members of the Lawther Committee, the report and its conclusions were brutally attacked (Extracts 5.7 and 5.8). As two of them subsequently wrote:

> *Nothing in our previous training or experience had prepared us for the experience of trying to evaluate scientific evidence against the background of populist pressure and in some cases, personal attack. Scientists take pride in seeing all sides of the argument; journalists want to communicate simple conclusions devoid of caveats and to sell papers through eye-catching headlines.*

These are the words of people who had not fully grasped the nature of the arena that they had agreed to enter. The mood of an especially vociferous section of spectators is captured by Figure 5.2. In 1980, there had been nearly a decade of environmental agitation about leaded petrol in the UK. This had produced an atmosphere in which what was wanted was not scientific detachment, but a judgement couched in scientific language which, if it erred, made sure that it erred on the side of prudence. Looked at in this light, the Lawther Committee's conclusions were vulnerable: once prudence was set loose on them, it could run amok.

First of all, by the Committee's own admission, there was no evidence for a 'safe' threshold in blood lead concentration; why therefore choose $35\,\mu g\,dl^{-1}$ as a trigger for action? This after all was the *highest* of the limits used in the EEC Directive to specify an exposed population that needed attention. Would it not have been prudent to use a lower figure? Hindsight if anything, strengthens this point because, as you know from

Extract 5.7 From *The Times*, 31 March 1980.

Lead pollution

From Dr R. Russell Jones

Sir, Eminent scientists who find themselves defending outmoded concepts exhibit a depressing obsession with methodology and issue unrealistic demands for incontrovertible proof. Professor Lawther's working party report on lead pollution (report, March 29) is a classic illustration of this phenomenon.

The accumulated evidence on this subject is already so disturbing that it becomes morally indefensible to sacrifice further generations of urban children in the hope that absolute proof will one day become available. The Committee's only contribution to the problem has been to allow another year to pass without effective action being taken. Let us now pray that the Government is not misled by their complacency.

Yours faithfully,
R. RUSSELL JONES
115 Moore Park Road, SW6.
March 30.

Extract 5.8 From *The Times*, 2 April 1980.

Health risks from lead

From Lord Ashby, FRS

Sir, You write in your leader (March 29) that the report on *Lead and Health* 'will not satisfy either side'. This is a disquieting statement. Scientific research is not done to satisfy pressure groups; it is done to ascertain the truth so far as that is possible. When the public take sides on a highly emotive issue, truth is the first casualty. Thanks to the courage of Professor Lawther and other scientific workers, willing to publish the truth as they see it, the Government now knows more about lead and health than it ever has before.

The thanks these scientific workers get for telling the truth is to be branded as 'complacent', to be accused of making a 'cover-up', of producing a 'political document' and (*The Times*, March 31) of 'defending outmoded concepts'. If integrity in reporting scientific work is an outmoded concept, it's a poor prospect for Britain.

Careful work published by the Department of the Environment, the Atomic Energy Research Establishment, and now the Medical Research Council, demonstrates (i) that lead levels in some places are too close to danger level to be tolerated; (ii) that the greatest risk to children from lead comes from water supplies in lead pipes and from paint; (iii) that lead from car exhausts contributes to lead in the atmosphere but is by no means the most dangerous source. These facts are not disputed by anyone who has taken the trouble to read the evidence. They are essential for the political decision the Government has to make. Pressure groups are free, of course, to use the facts; but not to distort them, to lie about them, or to impugn the integrity of the scientists who established them.

Since anyone who writes about lead levels is under suspicion of being on one 'side' or the other, I had better add that I am a pensioner: I have no interests in the oil lobby or the transport lobby; I use a bicycle and if I were asked how best to protect children from automobiles, I would suggest that the money saved by not taking lead out of petrol should be spent on doing something to prevent some 4 000 pedestrian children under 10 being killed or seriously injured on the roads of Britain every year.

Yours faithfully,
ASHBY,
House of Lords.

Still pumping poison into our children

Drawing by Peter Clarke

There is no longer any argument about the danger of lead pollution. But, argues Robin Russell Jones, commercial interest has once again triumphed over public safety, for the arguments used to keep lead in petrol are the same as those used 70 years ago to keep lead in paint

How much more evidence is needed before the Government takes effective action to abate this appalling menace?

Figure 5.2 An example of a newspaper article which, in its headlines and illustration, makes no concessions to alternative points of view. From *The Guardian,* 1 October 1981.

Section 2.4.1, in 1982 a level of 25 $\mu g\,dl^{-1}$ became the recommended trigger for action. Again, if Needleman's work provided 'some evidence of an association between raised tooth dentine levels and a slight lowering of measured intelligence', might it not be prudent to accept a causal link between raised lead levels and lowered IQ as a working hypothesis, at least until the further studies that the committee called for clarified the position. There was also much subsequent discussion of the precise meaning of the Committee's recommendation that emissions to the air of lead from traffic should be 'progressively reduced'. Members were to subsequently claim that this was a call for an ultimate reduction to zero. If so, they were guilty of a damaging lack of clarity. Their report contained no unequivocal recommendation of lead-free petrol, and consequently the government felt free to adopt the second of the three options described at the beginning of Section 5.2. On 11 May 1981, Tom King, the Secretary of State for the Environment, announced that from January 1986 the lead content of petrol would be limited to 0.15 $g\,l^{-1}$.

5.4 The environmentalists regroup

Despite its disappointment, the environmental lobby now rallied and changed its structure. CALIP was absorbed into the larger organization, CLEAR (Section 1.3), which was publicly launched on 25 January 1982 (Extract 5.9). The time was propitious. There was now stronger evidence, from the Turin experiment and other work, that the Lawther Committee's 10% estimate of the contribution of leaded petrol to blood lead was too low. Moreover, two of the Committee's members, Dr Yule and Dr Lansdown, had carried out a study (study 1 of Table 4.4) of the behaviour, intelligence and blood lead levels of London children which gave results similar to Needleman's in the blood lead range 7–32 $\mu g\,dl^{-1}$. This range fell below the Committee's threshold for remedial action of 35 $\mu g\,dl^{-1}$. Within a fortnight of its launch, CLEAR was able to pass to *The Times* a copy of a letter leaked to it from within the DHSS (Extract 5.10). This letter, written by the Department's Chief Medical Officer, Sir Henry Yellowlees, in March 1981, accepted that 'there is a strong likelihood that lead in petrol is permanently reducing the IQ of many of our children' and advised 'that action should now be taken to reduce markedly the lead content of petrol'.

The extent to which 1982 provided the environmentalists with a window of opportunity is revealed by Table 4.4. The first British study of 1981 had given marked support to CLEAR, but three further surveys were soon to be published, in 1983, 1984 and 1986, and all were to be unfavourable. Moreover, one of them, study 4, was car ried out by Lansdown and Yule, the authors of study 1 which featured strongly in the environmentalist campaign (Extract 5.11).

Extract 5.9 From *The Times*, 25 January 1982.

New fight on lead in the air

By Pearce Wright
Science Editor

More than 100 MPs are among supporters of the aims of the Campaign for Lead-free Air, CLEAR, which is to be launched today.

The new campaign is based on evidence to be presented today which is said to underline the argument that tiny amounts of lead in the air at present regarded as safe, can cause brain damage in children.

The group is to set up a trust supervised by people who include Dame Elizabeth Ackroyd, Dr Jonathan Miller, Lord Avebury, the Bishop of Birmingham and Mr Clive Jenkins.

An equally impressive list of professional people form CLEAR's medical and scientific advisory board.

Extract 5.10 From *The Times*, 8 February 1982.

Petrol: must our children still be poisoned?

by Des Wilson

I was angered all last week after being handed a confidential letter by the nation's top medical adviser to senior Whitehall colleagues which warned in uninhibited language of the danger to children from lead in petrol. Had the letter been made public at the time that it was written, we would now be on our way to lead-free petrol and our CLEAR (the Campaign for Lead-free Air) campaign would never have been necessary.

Extract 5.11 From *The Guardian*, 1 October 1981.

Although Needleman's findings received considerable publicity, the implications of his work are seldom discussed. If lead is a major factor in the development of abnormal behaviour and educational underachievement, then no child brought up in an urban environment can be considered immune from its effects. Dr Yule and Dr Lansdown have provided confirmation that this is so.

In May 1982, CLEAR organized an academic conference in London on the theme *Lead versus Health*. Patterson, Needleman, Lansdown and Yule all contributed papers. The conference chairman was Professor Michael Rutter, who had been a member of the Lawther Committee, and his summing up showed what inroads the environmentalist case had made since the publication of the Lawther report. Rutter concluded that the Lawther estimate of 10% for the total contribution of petrol to a person's lead uptake was too low: a figure of at least 25% was probably more appropriate. He noted stronger evidence for the hypothesis that quite low levels of lead exposure led to psychological impairment, and issued an appropriately prudent judgement: 'It would be both safer in practice, and scientifically more appropriate, to act as if the hypothesis were true rather than to do nothing on the assumption that it might be false'. Rutter noted that if the effect existed, it was small—deficits of 3–5 points in IQ—and dismissed claims that much delinquency and education retardation are caused by lead exposure. However, the costs of removing lead from petrol were 'quite modest by any reasonable standard', and removal would be 'a worthwhile and safe public health action'.

From an environmentalist's standpoint, the conference had been a triumph. A judicious mixture of disinterested scientists with others who were deeply committed to the anti-lead cause had yielded a collection of contributions, capped by an authoritative summing up, that was favourable to CLEAR's aims. The conference also provided welcome ammunition for Des Wilson, the chairman of CLEAR, who was now able to present the organization's testament in a book entitled *The Lead Scandal; the Fight to Save Children from Damage by Lead in Petrol.*

A crucial consequence of the agitation created by CALIP and CLEAR was the decision of the Royal Commission on Environmental Pollution to devote its 9th Report to *Lead in the Environment*. This decision was announced in March 1982, and the report was published in early 1983. Something of its flavour was conveyed in Chapter 4. This time prudence triumphed, as three of the Report's recommendations make clear:

> *The reduction of the maximum permitted lead content of petrol to 0.15 g/l should be regarded as an intermediate stage in the phasing out of lead additives altogether.* (para. 9.26)

> *The Government should begin urgent discussions with the UK oil and motor industries in order to agree a timetable for the introduction of unleaded petrol, having regard to the time required for essential production changes and the desirability of matching major refinery investment to long-term rather than interim requirements.* (para. 9.28)

> *The price of unleaded petrol should not exceed that of the highest grade of leaded petrol during the period in which leaded petrol is phased out.* (para. 9.29)

Today, oil companies sell unleaded petrol of greater than 95 RON on almost every garage forecourt in Western Europe. This has been achieved by a variety of measures. In Germany and Sweden for example, refining practices have been changed to increase the proportion of aromatics to about 40%. In the UK, this proportion is increasing as well, but in part, knock-resistance has been sustained by blending in alkanes of low molecular mass such as butane, $CH_3CH_2CH_2CH_3$. Because of its low molecular mass, the boiling temperature of butane is low, and this increases the volatility of the petrol, making loss by evaporation more likely; but as Table 3.2 shows, the octane number of straight-chain alkanes increases rapidly as the number of carbon atoms decreases, and the value for butane (113 RON) has a very favourable effect on knock resistance. Other measures include the use of oxygenates, (at levels up to about 5%), the lowering of compression ratios, and other changes to engine design.

Summary of Chapter 5

No explicit summary is given for Chapter 5. Instead, you will carry out a summarizing process of your own by doing Activity 5.1. This activity also provides you with an opportunity to practice higher-level skills concerned with the organization of information, the recognition of the limitations of the scientific attitude, and with written communication. Although you can refer to the newspaper extracts if you wish, the sample answers have been written without reference to them.

Activity 5.1 *You should spend up to 1 hour on this activity.*

Chapter 4 describes five different areas of scientific work on environmental lead, and gives a response of the Royal Commission to each one. Chapter 5, on the other hand, gives an impression of the relative prominence given to the different areas in the political campaign. Was the scientific area most prominent in the campaign, the one which had most impact on the Royal Commission? If you think not, suggest a reason for the difference.

A possible procedure

If this kind of exercise is new to you, you should find the following suggestions useful:

(i) First list the five areas of scientific work, and then assess the responses of the Royal Commission to them. Do this by reading the quotations from Commission reports, at the end of Sections 4.1, 4.2.3, 4.3.2, 4.4 and 4.5.2. Try to judge, using these quotations, which issue was the most influential, and which the least.

(ii) You have already read Chapter 5; now glance quickly over it again, noting points at which the five areas receive specific or implied mentions. As noted above, you are not expected to read the newspaper extracts during this exercise. This skim-reading approach is valuable because it quickly eliminates material like general chronology which is not relevant to the activity. Now carefully re-read the references that you spotted, and select the scientific area that you think was most important in the political campaign. Make notes explaining why the contents of Chapter 5 justify your choice, rather than the choice of other areas that are mentioned.

(iii) Now compare the choices that you made in (i) and (ii). If the choice that you made under (ii) differs from that made under (i), think carefully about this preference of the political campaigners, and note possible reasons for it.

(iv) Now write a three-paragraph answer to the activity, devoting one paragraph to each of items (i), (ii) and (iii) above. Your total answer should be no more than 450 words: the answer given on pp. 85–86 is less than 400.

---

6 Conclusion

The environmentalist campaign against leaded petrol was very effectively conducted, and ultimately triumphant. But the introduction to this book gave notice that we would be alert to things about leaded petrol which eased the going for the environmentalist lobby. Eight relevant points are given below; most emerged naturally in the text, but some are new or, at least, are only now made explicit.

1 Lead is a well established poison: that is, there are many well documented cases of deaths caused by the inadvertent ingestion of quite small quantities of lead compounds (Section 2.2).

2 Blood lead levels in the population as a whole are closer to officially recognized danger levels than for any other toxin (Section 4.1).

3 Once there was official recognition of levels at which obvious symptoms of lead poisoning might appear (e.g. blood lead concentrations of $70\,\mu g\,dl^{-1}$), it was a plausible hypothesis that levels well below this might have subtle effects on behaviour which, although hard to detect, are nevertheless real (Sections 2.4 and 4.3).

4 It was possible to present children as being particularly at risk, and children have a huge emotional claim on the nation's consideration (Section 4.3).

5 Background lead in the environment is primarily anthropogenic. This offered greater opportunities for generating public guilt about environmental lead, and guilt is a spur to action (Section 4.2). There has, for example, been no comparable agitation (until recently) about the protection of the population from emissions of background, natural radon gas which are believed to cause some 2500 UK deaths per year.

6 In the UK, lead alkyls were made by Associated Octel, which was owned by a consortium of multinational oil companies. The major car manufacturers which built vehicles to run on leaded petrol are also multinationals. The environmentalist campaign could therefore be staged as a contest with big business, in which the opposition's arguments were prejudiced by a desire for profit. This attracted support from important sections of political opinion in the UK that are hostile to large companies. Multinational companies are peculiarly vulnerable in this respect although, or perhaps because, attitudes towards them are ambivalent. Both Labour and Conservative governments have been anxious to draw multinational investment into the UK, and offered generous concessions as bait. Yet once established, multinationals can be branded as unpatriotic or immoral because their interests transcend national boundaries, and they can mitigate the effects of a disagreeable national policy in one country by moving resources to another.

7 The removal of lead from petrol makes possible the reduction of other unpleasant pollutants, such as nitrogen oxides, carbon monoxide and hydrocarbons, through catalytic conversion (Section 4.5).

8 There were obvious solutions to the problem, and the social costs, such as unemployment and higher prices, were not enormous (Section 5.2).

Activity 6.1 *You should spend up to 1 hour on this activity.*

In this final activity, skills that were practiced on the book text in Activity 5.1 are applied to the less familiar medium of a published scientific article (Extract 6.1). The article was published in 1971 by Professor Derek Bryce-Smith in the journal *Chemistry in Britain*, and its appearance marked the beginnings of the organized UK campaign against leaded petrol. Do not be put off if you find large parts of the extract unintelligible. Scientists are very often in this position when they read scientific papers: what they do is to extract information from those parts of the paper that they *do* understand, and indeed, perhaps understand better than the paper's author.

(a) Chapter 4 identified five areas of scientific study that, in 1982–3, influenced the Royal Commission's judgement against leaded petrol. How many of the five are referred to in this 1971 article? (Skim-read Extract 6.1 until you detect a reference to one of the five areas. Read that part carefully and mark or note the sentence which is the most obvious or forceful mention of the scientific area in question. Repeat this process until you finish the extract. Now list the areas which have been mentioned, selecting a sentence to represent each one.)

(b) There is a broad and important area of scientific study about the effects of lead which is referred to in Extract 6.1, but which has not been mentioned in the book. What is it? (The exercise completed in (a) may have revealed the answer. If not, question (c) may give you a clue.)

(c) Chapter 2 stated that symptoms of lead poisoning included pallor and anaemia. Select a sentence from Extract 6.1 which could be used in an explanation of this.

(d) In Chapter 2, you were warned that unusual units often occur in scientific papers. Bryce-Smith expresses blood lead concentrations in p.p.m. This means parts per million, for example, 5 p.p.m. means 5 grams of lead in one million grams of blood. Assuming that the density of blood, like that of water, is $100\,g\,dl^{-1}$, describe how the figures in the lower part of Figure 2 in Extract 6.1 should be changed so that the units are the same as those used throughout this book.

Extract 6.1 From *Chemistry in Britain* (1971) 7, pp. 54–56.

Lead pollution— a growing hazard to public health

D Bryce-Smith

Lead is ubiquitous: it is in the food we eat, the water we drink and the air we breathe. Yet it is one of the most insidiously toxic of the heavy metals to which we are exposed, particularly in its ability to accumulate in the body and to damage the central nervous system, including the brain. This article draws attention to some recent evidence that blood lead levels among the general population approach close to, and even exceed, those at which interference with metabolic processes can occur; and points out that a major cause of these high body burdens appears to be the inhalation of airborne lead resulting from combustion of the lead alkyl antiknock agents in petrol.

The toxicology of lead is very complex. Inorganic lead (Pb^{2+}) is a general metabolic poison and is cumulative in man. It inhibits enzyme systems necessary for the formation of haemoglobin through its strong interaction with –SH groups, and has been said to interfere with practically any life-process one chooses to study.[1] Children and young people appear specially liable to suffer more or less permanent brain damage, leading *inter alia* to mental retardation, irritability and bizarre behaviour patterns. Lead-induced psychosis is said to show striking similarities to the manic-depressive type,[1a] Pb^{2+} can replace Ca^{2+} in bone, so tends to accumulate there; but an unpleasant feature is that it may be remobilized long after the initial absorption, *e.g.* under conditions of abnormally high calcium metabolism such as feverish illness, during cortisone therapy, and also in old age. It can cross the placental barrier, and thereby enter the foetus.

Lead alkyls such as tetraethyl-lead are even more poisonous than Pb^{2+}, and are handled quite differently in the body: the effects are probably due to the derived R_3Pb^+ species. Urban atmospheres can contain *ca* 2 per cent—at most 10 per cent—of their lead in this 'organic' form.[1] The earliest toxic symptons which, one must emphasize, have been observed in men and experimental animals at air concentrations much higher than those now normally found in cities, are psychical, *e.g.* excitement, depression, irritability. More serious occupational exposure can lead to insanity and death: cases have been recently reported from Greece.[2] Oddly enough, it is not apparently known whether children are more sensitive than adults to lead alkyls; but experience with the related alkylmercury compounds,[3] and with Pb^{2+},[1] suggests that it would be prudent to assume greater sensitivity in the absence of contrary evidence. It therefore remains to be established whether the present environmental exposure of children to 'organic' lead is safe or harmful.

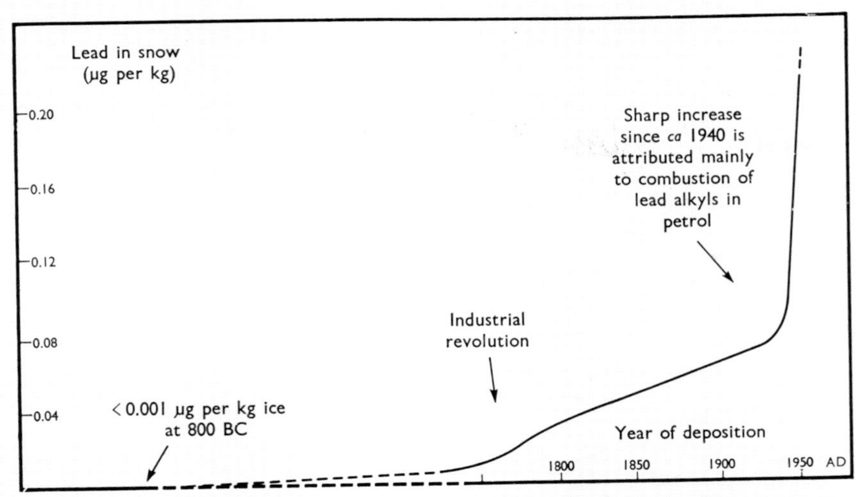

Fig. 1. Lead content of snow layers (→ice) in Northern Greenland.[4] Lead levels in South Polar icesheets were too low for detection before 1940, but have now risen to about 0.02µg per kg.

Man has evolved in the presence of a certain amount of inorganic lead from natural sources, to which he might have adapted, so any assessment of the threat from environmental pollution must consider the increase above natural background levels during biologically recent time. To begin with, we know that over 2000kt of lead is mined each year, in comparison with about 180kt estimated to be naturally mobilized and discharged into the oceans from rivers. A recent examination by Japanese and American workers of annual snow strata in N. Greenland has revealed most elegantly that levels of airborne lead have increased markedly since the Industrial Revolution, and very sharply since about 1940: the findings are illustrated in *Fig. 1*.[4] The increase in airborne lead is particularly important because *ca* 50 per cent of this lead can be absorbed on inhalation, in contrast with the 5–10 per cent absorbed from lead-contaminated food and water through the alimentary tract.* (Of course, fall-out of airborne lead also adds to the lead present in food and water.[1])

Several studies have shown that the principal source of airborne lead in urban environments is the lead-containing aerosol emitted with exhaust fumes from motor vehicles running on petrol containing lead alkyl antiknock agents, *see e.g.* refs 1, 5, 6.

The question remains whether lead is being absorbed by man from the environment in amounts which could hazard public health. And one must remember that the lead absorbed in seemingly harmless trace amounts over a long period can accumulate in the body to levels which exceed the 'threshold level' for potential poisoning, and thereby produce long delayed toxic symptons.[7] Even the present *average* blood lead levels of adults in industrialized countries are not far below those which can lead to overt clinical symptons of poisoning: and it has been authoritatively reported that actual levels are related *inter alia* to the opportunity for exposure to the exhaust fumes from motor vehicles.[6] *Figure 2* illustrates these points by reference to the variously termed threshold levels for potential poisoning which have been given by different workers. For comparison, the present body fat levels of DDT (plus its metabolic product DDE) are also shown.

It is clear from *Fig. 2* that present body burdens of lead pose a far greater threat to human health than those of DDT. And to place the effect of lead in even better perspective, to my best knowledge no other toxic chemical pollutant has accumulated in man to average levels so close to the threshold for overt clinical poisoning.

The results illustrated in *Fig.* 2 were obtained in the United States.[6] No comparable study with lead seems to have been made in Britain, although the levels of atmospheric lead in London appear to be of the same order as those found for comparable areas in the United States and elsewhere,[12] and lead levels in Warrington (Lancs), for example, are reported to be even higher than those found in California.[13] A recent survey from Liverpool University of lead in public water supplies showed that of 47 samples, 22 had lead levels at or above the internationally recommended maximum, and 3 had twice this amount.[14] The authors remarked that '. . . two million people were being regularly exposed to the risk of lead poisoning' from lead in water supplies.'†

Examination of a control group of Manchester children showed an average blood lead level of 0.309 ppm,[15] slightly higher than the average figures for American adults illustrated in *Fig. 2*, and 17 per cent of these children were reported to have levels above 0.5 ppm.‡ Thus there is evidence that lead pollution is at least as serious a problem in Britain as in comparable regions in the United States.

Within the last year, Hernberg and Nikkanen (Helsinki)[16] and Millar *et al.* (Glasgow)[17] have independently presented most significant evidence that enzyme inhibition by inorganic lead occurs even over the range 0.2–0.4 ppm in blood, *i.e.* at the levels now present among the general urban population (*Fig. 2*). It follows that most people in towns and cities can already expect to be undergoing some interference with metabolism through their absorption of lead. Indeed, Danielson[1] has recently estimated that present levels of inorganic lead absorption are 5–10 times higher than that

* This difference in absorption has led to confusion about the relative importance of airborne and dietary lead. Typically, an urban adult breathing air containing, say, 3µg m^{-3} of lead and respiring about 15 m^3 of air per day will *absorb* 20–25µg of lead per day from this source. Roughly the same amount (15–30µg) will be absorbed from a normal daily diet containing *ca.* 300µg of lead.

† Lead in drinking water appears to come partly from the still-continuing use of lead water pipes and soldered joints, and partly from atmospheric fall-out. A new reservoir to serve the London area is now under construction at Datchett on a site adjacent to the M4 motorway.

‡ The authors of ref. 9 reported that nearly half of a group of mentally retarded children had blood lead levels above the maximum level of 0.36 ppm found in a control group of normal children; see also similar work in ref. 18. Another recent study[17] found a much smaller proportion of raised blood lead levels in a group of mentally retarded children. The authors wrote that 'even modest elevations of blood lead may be associated with biochemical abnormalities in the child brain'.[17] Although a proportion of the raised lead levels may result from pica, *e.g.* the gnawing of paint, there is abundant clinical evidence that children are more prone than adults to suffer lead poisoning, with damage to the central nervous system: *see e.g.* refs 19, 20.

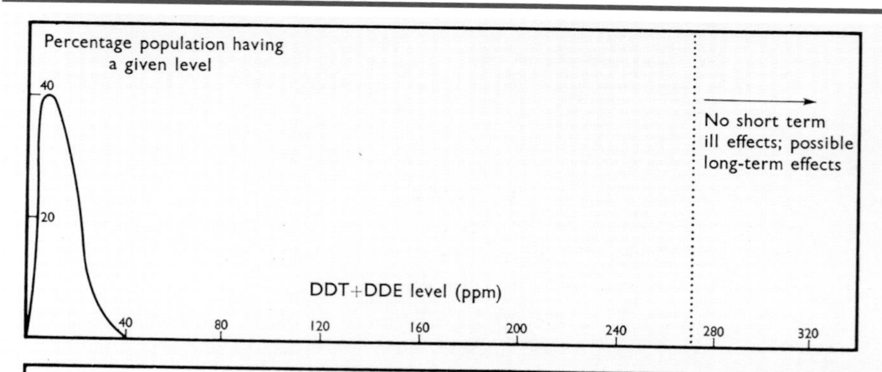

Fig. 2. Relation between accumulated pollutant levels in man and threshold levels for potential poisoning: (*left*) levels of DDT + DDE in body fat;[8] (*below*) lead levels in blood for Philadelphia males 1961–2—dashed line for suburban inhabitants and full line for downtown inhabitants.[6] 'Kehoes' toxic threshold figure of 0.8 ppm[11] is considered in ref. 1 to be inapplicable to the question of public health in cities. Moncrieff's threshold level of 0.36 ppm[9] is based on studies with children.

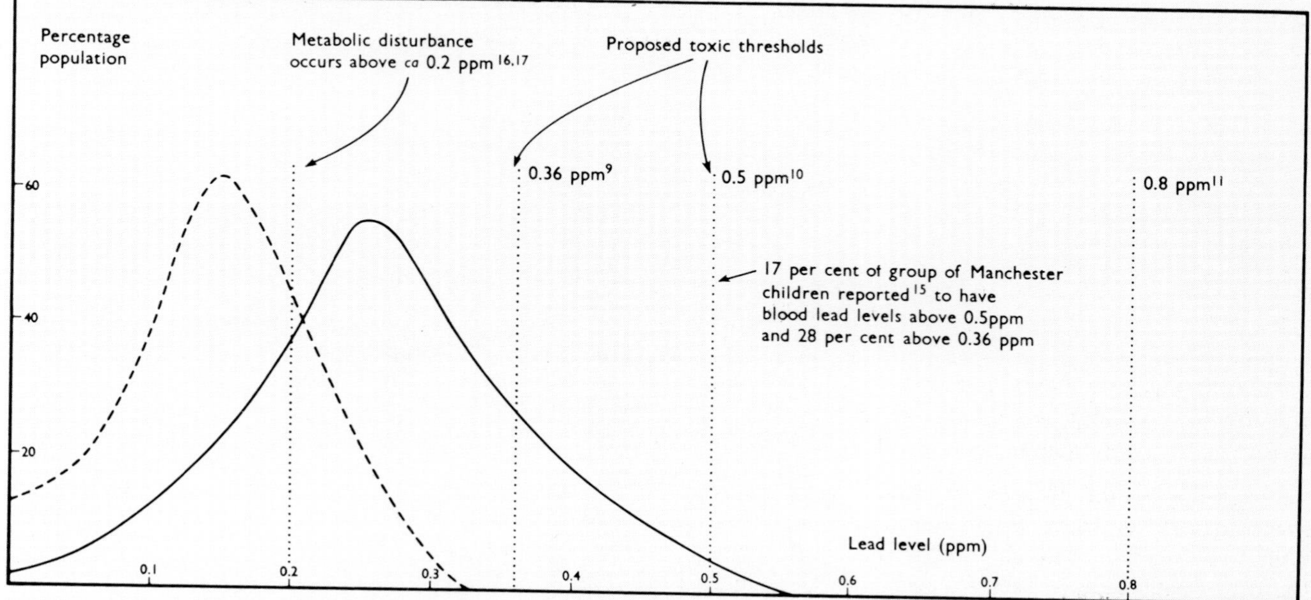

which might be toxicologically acceptable.

Legislative and economic aspects

Legislation designed to reduce the amounts of lead alkyls added to petrol is in force or under high level discussion in the United States, Japan, France, W. Germany and Sweden. United States government cars are required to use no-lead or low-lead petrol. The use of leaded petrol is prohibited in Moscow. The Soviet Union has set a maximum permissible figure of 0.7μg m^{-3} of lead for the general atmosphere: most of the values measured during a prolonged survey of air pollution in the City of London were higher than this, ranging up to 4.8μg m^{-3}.

There is no legislation in Britain to control the amount of lead which may be added to petrol or present in the general atmosphere. In fact, the British Standard Specification concerned with petrol (BSS 4040) was recently revised, but no change in the lead content was recommended. The extra cost of producing lead-free petrol would be ca 2d per gallon, possibly less,[21] and Shell and British Petroleum have separately announced recently that they are ready to provide such petrol for sale in Britain, as and when required. I venture to hope that the companies concerned will soon feel encouraged to implement their public-spirited offer.

References

1. L. Danielson, *Bulletin No.* 6, *Ecological Research Committee*. Stockholm: Swedish Natural Science Research Council, 1970: see also refs. therein.

1*a*. H. Stern, *Chicago med. Sch. Q.*, 1969, **3**, 87.

2. *The Times*, 7 December 1968.
3. Reviewed by G. Löfroth, *Bulletin No.* 4, *Ecological Research Committee*. Stockholm: Swedish Natural Science Research Council, 1969.
4. M. Murozumi, T. J. Chow and C. C. Patterson, *Geochim. cosmochim. Acta*, 1969, **33**, 1247.
5. T. J. Chow, *Nature, Lond.*, 1970, **225**, 295, and references therein: see also *Chem. engng News*, 9 March 1970, p. 42.
6. J. H. Ludwig, D. R. Diggs, H. E. Hesselberg and J. A. Maga, *Am. ind. Hyg. Ass. J.*, 1965, **26**, 270.
7. E. Browning, *Toxicity of industrial metals*, 149. London: Butterworths, 1961.
8. K. A. Hassall (University of Reading) kindly provided this graph.
9. A. A. Moncrieff, O. P. Koumides, B. E. Clayton, A. D. Patrick, A. G. C. Renwick and G. E. Roberts, *Archs Dis. Childh.*, 1964, **39**, 1: *cf. Br. Med. J.*, 1967, **3**, 174.
10. R. Egli, E. Grandjean, J. Marmet and H. Kapp, *Schweiz. med. Wschr.*, 1957, **87**, 1171.
11. R. A. Kehoe, *J. R. Inst. publ. Hlth Hyg.*, 1961, **24**, 177.
12. R. E. Waller, B. T. Commins and P. J. Lowther, *Br. J. ind. Med.*, 1965, **22**, 128.
13. E. Ward (Chief Public Health Inspector, Warrington), quoted in *The Guardian*, 15 May 1970.
14. J. A. Tolley and C. D. Reed, *Ecologist*, 1970, **1** (4), 31.
15. N. Gordon, E. King and R. I. Mackay, *Br. Med. J.*, 1967, **2**, 480.
16. S. Hernberg and J. Nikkanen, *Lancet*, 10 January 1970, p 63.
17. J. A. Miller, V. Battistini, R. L. C. Cumming, F. Carswell and A. Goldberg, *Lancet*, 3 October 1970, p 695.
18. S. L. M. Gibson, C. N. Lam, W. M. McCrae and A. Goldberg, *Archs Dis. Childh.*, 1967, **42**, 573.
19. E. M. Rathus, *Med. J. Aust.*, 1967, **19**, 371.
20. M. A. Perlstein and R. Attala, *Clin. Pediat.*, 1966, **5**, 292.
21. *Manufacture of unleaded gasoline*, US motor gasoline economics, vol. 1. Houston, Texas: Bonner and Moore Associates, 1967.

7 Reverberations

This short section provides some information on scientific work that has been carried out since the legislation on leaded petrol was implemented. Surveys by the Department of the Environment have shown that blood lead levels in the UK have fallen since 1984. For a sample of about 400 inner-city men, the most frequently observed value in 1984 was $13\,\mu g\,dl^{-1}$; by 1987, the figure was $10\,\mu g\,dl^{-1}$. The corresponding values obtained for a sample of nearly 1 000 children were $10\,\mu g\,dl^{-1}$ and $7\,\mu g\,dl^{-1}$. More recent data suggest that such falls continued up to 1991. The reductions may be due not just to the decline in the use of leaded petrol, but to other steps which have been taken to cut lead uptake from tap-water, food and paint.

In 1991, new measurements of the concentration of lead in Greenland snow corroborated Patterson's values (Section 4.2.2), and supported his claim that there had been a roughly 200-fold increase between prehistoric times and 1965. However, the more recent measurements also showed that between 1969 and 1989, the concentrations had fallen back to about one-eighth of their 1960s levels (see Figure 7.1).

The fall from about 1970 to 1989 was attributed to the decline in the use of leaded petrol across the Northern Hemisphere. The full publication is provided as an offprint, but you are not expected to read it unless asked to do so as part of an assignment. It is interesting that the measurements were made by atomic absorption spectrometry, a technique whose accuracy has increased substantially since the 1960s when Patterson's measurements were made by isotope dilution analysis.

Although the decision to phase out leaded petrol has reduced the prominence of environmental lead in the newspapers and on television, many scientists continue to work on its measurement, and on the effects that it has on both children and adults. The list of publications which follows provides a good survey some of the more recent work.

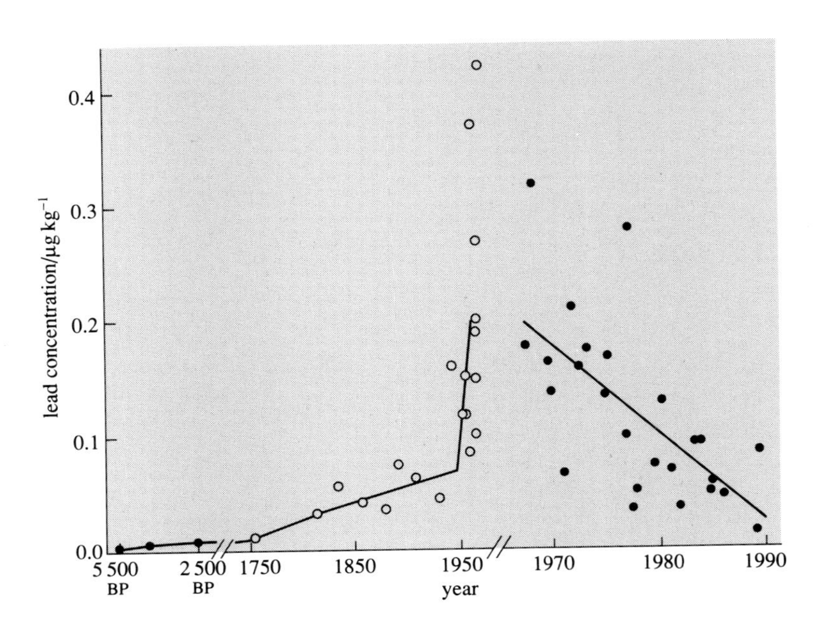

Figure 7.1 This shows how the trend evident in Patterson's plot of Arctic lead levels against time (Figure 4.5) has been reversed since the late 1960s. The new data are shown by filled circles; Patterson's data, by open circles. By 1990, lead concentrations in Arctic snows had fallen back towards their prehistoric levels, being only about one-eighth of the values observed for the 1960s. (BP = before present.)

Further reading

Should you be interested in obtaining more detailed information about some of the topics discussed in this book, the following sources are helpful:

Royal Commission on Environmental Pollution (1983), 9th Report, *Lead in the Environment*, HMSO, London.
This is an invaluable source of information. Its tone is neutral in relation to interested parties, but it also takes account of political realities.

Wilson, Des (1983) *The Lead Scandal*, Heinemann Educational Books.
A well written, strident environmentalist text with much interesting detail of the political machinations behind the leaded petrol controversy.

Rutter, M. and Jones, R. R. (eds) (1983) *Lead versus Health*, John Wiley & Sons, New York.
This contains the works presented at the symposium organized by CLEAR in May 1982. It contains papers from Patterson, Needleman, Lansdown, Yule and Rutter.

Hancock, E. G. (ed) (1986) *Technology of Gasoline*, Blackwell Scientific Publications.
A clear and concise account of both the history and current manufacture of petrol.

Lansdown, R. and Yule, W. (1986) *The Lead Debate*, Croom Helm, London.
An excellent, detached account of lead in the environment and of its influence on child health.

Fergusson, J. E. (1990) *The Heavy Elements: Chemistry, Environmental Impact and Health Effects*, Pergamon, Oxford.
Extensive parts of this work deal with lead and the environment; it mentions some of the more recent work on the subject.

Fulton, M. *et al.*, (1987) *The Lancet*, no. 8544, pp. 1221–1226.
This gives the results of the largest and most recent study of the effects of lead on child attainment in Britain: study 5 in Table 4.4.

Skills

In the *Course Study Guide*, you will find a list of the skills which *Science Matters* will impart. They are phrased in very general terms, and might better be described as skill categories. If you read through them, you should recognize that, in this book, you have done things that fall within their domain. However, not all of those things have been *explicitly* taught: you have done them in passing when your main purpose was something else—for example, while trying to understand the science of the book.

In this section, therefore, we list skills that *have*, in some sense, been explicitly taught and/or revised. You should find that most of them are special instances of the general skill categories, although some, such as 5 and 6, are deeply rooted in the particular content of this book. After each one, there is a list of questions and activities where that skill is practised.

1 Add, subtract, multiply and divide numbers in scientific notation. (*Questions 1.1, 1.2, 2.1, and 4.1*)

2 Convert scientific quantities from one set of units to another. (*Question 2.4; Activity 2.3*)

3 Interpret data presented in different forms (e.g. tables, graphs, histograms, bar charts and diagrams). (*Questions 2.3 and 4.1; Activities 2.1, 2.2, 4.2, 4.4 and 6.1*)

4 Show that changes in the way that data is visually presented may change the impression that the data conveys. (*Activity 4.1*)

5 Balance chemical equations. (*Questions 1.4 and 2.2*)

6 Use and show familiarity with structural formulae for molecular compounds, recognizing the relationship of those formulae to the number of bonds that each type of atom forms. (*Questions 3.1, 3.2 and 3.3*)

7 Distinguish between correlations and causal relationships. (*Question 4.2; Activity 4.3*)

8 Extract from a section of teaching text that you have studied, information that is relevant to a particular problem, and then use your understanding of the text to combine that information to frame an answer to the problem, which you then write out in your own words. (*Activities 3.1, 3.2 and 5.1*)

9 Extract from a scientific article, large parts of which you may find unintelligible, information that is relevant to a particular problem, and by integrating that information with what you already know, give an answer to the problem in your own words. (*Activity 6.1*)

Answers to questions

Question 1.1

(a) (i) 4.2×10^{8} kg.

Starting with a figure of 4.2, the decimal point must be advanced eight places to give 420 000 000.

(ii) 2.3×10^{-7} m.

Starting with a figure of 2.3, the decimal point must retreat seven places to give 0.000 000 23.

(b) (i) 360 000 kg.

The figure 3.6×10^{5} tells us that if we start with 3.6, the decimal point must be advanced five places over the six and the necessary added zeros to give the number in ordinary decimal form, 360 000 kg.

(ii) 0.000 085 m.

Likewise, the figure 8.5×10^{-5} tells us that the decimal point in the number 8.5 must retreat five places to give 8.5×10^{-5} in decimal form, 0.000 085 m.

Question 1.2

(a) 8.4×10^{14} kg m s^{-1}.

The first figure comes from the multiplication 4.2×2.0, the second from the multiplication $10^{8} \times 10^{6}$. Note that on multiplication the powers of the tens are *added*. Note also that units must be included.

(b) 2.1×10^{2} m s^{-1}.

The first figure comes from the division, $4.2 \div 2.0$, the second from the division, $10^{8} \div 10^{6}$. Note that on division, the powers of the tens are *subtracted*. Note also that the *units* are divided along with the numbers: a distance (m) is divided by a time (s) to give a speed whose units are metres per second, or m s^{-1}:

$$\mathrm{m} \div \mathrm{s} = \frac{\mathrm{m}}{\mathrm{s}} = \mathrm{m} \times \frac{1}{\mathrm{s}} = \mathrm{m} \times \mathrm{s}^{-1} = \mathrm{m\,s}^{-1}$$

(c) 8.4×10^{2} m.

Again, on multiplication the powers of the tens are added: $10^{6} \times 10^{-4} = 10^{6+(-4)} = 10^{2}$. The units show that this is a multiplication of a speed (m s^{-1}) by a time(s) which yields a distance (m):

$$\mathrm{m\,s}^{-1} \times \mathrm{s} = \mathrm{m} \times \frac{1}{\mathrm{s}} \times \mathrm{s} = \mathrm{m}$$

(d) 2.1×10^{7} m s^{-2}.

Again, on division the powers of the tens are subtracted: $10^{3} \div 10^{-4} = 10^{3-(-4)} = 10^{7}$. The units show that this is a division of a speed (m s^{-1}) by a time (s) which yields an acceleration (m s^{-2}):

$$\mathrm{m\,s}^{-1} \div \mathrm{s} = \frac{\mathrm{m\,s}^{-1}}{\mathrm{s}} = \mathrm{m} \times \frac{1}{\mathrm{s}} \times \frac{1}{\mathrm{s}} = \frac{\mathrm{m}}{\mathrm{s}^{2}} = \mathrm{m\,s}^{-2}$$

Question 1.3

(a) $PbSO_4$ contains three different kinds of atom, symbolized Pb, S and O. These are lead, sulphur and oxygen atoms respectively. The subscript numbers against these atoms in the formula are lead, one (implicitly), sulphur, one (implicitly) and oxygen, four. As this is an empirical formula, this tells us that there is one sulphur atom and four oxygen atoms for every lead atom in lead sulphate.

(b) The formula $PbO_2(s)$ tells us that there are two oxygen atoms for every lead atom in lead dioxide; the formula $CO_2(g)$ tells us that there are two oxygen atoms for every carbon atom in carbon dioxide. Thus far, the two formulae are equally informative. But you were told that $CO_2(g)$ is a *molecular* formula and this tells you that the gas is composed of individual molecules containing one carbon atom and two oxygen atoms. In carbon dioxide gas, these molecules move about independently of one another. As $PbO_2(s)$ is only an *empirical* formula, you cannot assume that lead dioxide contains PbO_2 molecules, and in fact it does not: the lead and oxygen atoms are arranged in a more complicated way such that individual PbO_2 molecules cannot be picked out. Thus the *molecular* formula, $CO_2(g)$, carries more information.

Sometimes a chemical formula is quoted, and it is unclear whether its formula is empirical or molecular. If you assume on these occasions that the formula is empirical, you will draw no wrong conclusions. However, formulae cited for gases and liquids are *usually* molecular, and you will not often go wrong if you make this assumption.

Question 1.4

$$2PbO(s) + C(s) \longrightarrow 2Pb(l) + CO_2(g)$$

The equation is balanced because there are two lead atoms, two oxygen atoms and one carbon atom on each side of the equation, and because there is the same total charge (zero) on each side as well.

Question 2.1

$5.088 \times 10^{14}\,s^{-1}$.

Using Equation 2.5, with $\Delta E = 3.371 \times 10^{-19}\,J$, and $h = 6.626 \times 10^{-34}\,J\,s$,

$$\begin{aligned} 3.371 \times 10^{-19}\,J &= (6.626 \times 10^{-34}\,J\,s) \times f \\ \text{Therefore } f &= \frac{3.371 \times 10^{-19}\,J}{6.626 \times 10^{-34}\,J\,s} \\ &= 5.088 \times 10^{14}\,s^{-1} \end{aligned}$$

By convention, the unit of frequency, s^{-1}, which means 'per second', is known as the Hertz, Hz. Since 5.088×10^{14} is just over five hundred million million, the frequency of $5.088 \times 10^{14}\,Hz$ tells us that when this orange light travels through a vacuum, over 500 million million crests of the light wave pass a fixed point every second.

Question 2.2

(a) $PbS(s) + PbSO_4(s) \longrightarrow 2Pb(l) + 2SO_2(g)$

The equation has two lead atoms, two sulphur atoms and four oxygen atoms on each side of the equation.

(b) $PbCO_3(s) + 2H^+(aq) \longrightarrow Pb^{2+}(aq) + CO_2(g) + H_2O(l)$

The equation has one lead atom, one carbon atom, two hydrogen atoms and three oxygen atoms on each side of the equation. It is therefore balanced with respect to atoms. It is also balanced with respect to charge: the sum of the charges on each side of the equation is +2.

Question 2.3

(a) $0.6\,\mu g\,ml^{-1}$.

In Figure 2.10, the peak marked 'atomize' is the one whose height gives an absorbance due to lead and must be compared with the standard. (Compare the discussion of Figure 2.5.) The absorbance for the standard is 0.28, and for the workman's sample, it is 0.42. As absorbance is proportional to concentration under the same spectrometer conditions:

$$\frac{\text{lead concentration in sample}}{0.4\,\mu g\,ml^{-1}} = \frac{0.42}{0.28}$$

$$\text{Therefore lead concentration in sample} = \frac{0.42}{0.28} \times 0.4\,\mu g\,ml^{-1}$$

$$= 0.6\,\mu g\,ml^{-1}$$

Note that the sampling volume, $2\,\mu l$, is not needed for the calculation, except insofar as it tells us that the volumes of sample and standard were identical.

(b) The workman's blood lead concentration is $0.6\,\mu g\,ml^{-1}$. As there are 100 ml in 1 dl, this is equal to $60\,\mu g\,dl^{-1}$. This level does not necessitate suspension in the UK where the threshold is $70\,\mu g\,dl^{-1}$, but it would in the USA where the threshold is $50\,\mu g\,dl^{-1}$.

Question 2.4

(a) $5\,000\,\mu g\,l^{-1}$.

$$\text{Concentration of lead in cider} = 3 \text{ mg per pint}$$

$$= 3 \times \frac{\text{mg}}{\text{pint}}$$

In the given concentration, mg occurs on the top of the fraction, and pint on the bottom. Both these units must be eliminated. Consequently, the multiplying fractions must have mg on the bottom and pint on the top.

$$\text{As } 1 \text{ pint} = 0.576 \text{ litres}$$

$$\text{and } 1\,000\,\mu g = 1 \text{ mg}$$

The required forms are thus:

$$\frac{1 \text{ pint}}{0.576 \text{ l}} = 1$$

$$\frac{1\,000\,\mu g}{1 \text{ mg}} = 1$$

Multiplying our concentration by these two fractions:

$$\text{concentration of lead in cider} = 3 \times \frac{\text{mg}}{\text{pint}} \times \frac{1 \text{ pint}}{0.576 \text{ l}} \times \frac{1\,000\,\mu g}{1 \text{ mg}}$$

$$= \frac{3\,000\,\mu g}{0.576 \text{ l}}$$

$$= 5\,000\,\mu g\,l^{-1} \text{ (1 significant fig.)}$$

(b) This is 100 times the lead concentration recommended as an upper limit for tap-water in the EEC Directive.

Question 3.1

Structures XIV and XVII exist; the others do not, because in each case, rule 2 is violated. In structure XIII, the second carbon atom in the chain forms only three bonds, and in structure XV it forms five. In structure XVI, the topmost carbon atom in the ring also forms five bonds, three to carbon, and two to hydrogen atoms.

Question 3.2

The missing structures are XXI and XXII. Since all compounds with the molecular formula C_6H_{14} have the same molecular mass, their boiling temperatures should be similar. The boiling temperatures of the compounds shown as structures III–V are quoted in the text as 69 °C, 60 °C and 50 °C respectively, the average being 60 °C, so this is a reasonable estimate for XXI and XXII. The experimental values are 63 °C and 58 °C respectively.

```
      H  CH3 H
      |   |  |
H3C - C - C - C - CH3
      |   |  |
      H   H  H
```

XXI

```
      H3C CH3
       |   |
H3C - C - C - CH3
       |   |
       H   H
```

XXII

Question 3.3

The modified rules 1 and 2 have the form:

1 Each hydrogen atom forms just one bond which must link it to either a carbon or an oxygen atom.

2 Each carbon atom forms four bonds, any one of which may be linked to either a hydrogen atom, an oxygen atom, or another carbon atom.

The new rule, rule 3, has the form:

3 Each oxygen atom forms two bonds, one of which is linked to a carbon atom, and the other of which is linked to either a hydrogen, or to another carbon atom.

Question 3.4

The order is (iii) > (i) > (ii).
(iii) has the largest octane number because it is an aromatic compound, and (ii), being a straight-chain alkane has the least. Compound (i) is a branched-chain alkane and is intermediate in octane number.

Question 3.5

(i), (b); (ii), (c); (iii), (a); (iv), (d).
Hydrocarbons in petrol contain 5–10 carbon atoms, a property possessed by (i) and (iii). As (iii) is highly branched, it is a desirable component, but (i) is not because it is a straight-chain alkane. Catalytic reforming can, however, remedy this. Compound (ii) has too many carbon atoms, and too high a boiling temperature to be desirable, but can be broken down into suitable smaller molecules by catalytic cracking. Compound (iv) is methane, a gas.

Question 4.1

(a) $0.10\,\mu g\,kg^{-1}$.

If the mass of ^{207}Pb in the sample is x, then that of ^{208}Pb is $2.5x$, and after addition of the ^{208}Pb sample, the $^{208}Pb/^{207}Pb$ mass ratio becomes $(2.5x + 40\,\mu g)/x$. As before, we assume that the mass ratio is equal to the ratio of the peak heights, so:

$$\frac{(2.5x + 40\,\mu g)}{x} = \frac{130}{2}$$

By cross-multiplying, that is, by multiplying both sides by $2x$:

$$5x + 80\,\mu g = 130x$$

$$\text{Therefore } 125x = 80\,\mu g$$

$$\text{and } x = 0.64\,\mu g$$

As ^{207}Pb is about 21% of the total mass of lead in the 30 kg sample:

$$\text{Total mass of lead} = \frac{(0.64 \times 100)}{21}\,\mu g$$

$$= 3.05\,\mu g$$

$$\text{Thus concentration of lead} = \frac{3.05\,\mu g}{30\,kg}$$

$$= 0.10\,\mu g\,kg^{-1}$$

(b) The lead concentration in the ice is $0.10\,\mu g\,kg^{-1}$. From Figure 4.5, this ice was laid down fairly recently, the lines in the figure suggesting a date of about 1950.

Question 4.2

No, it does not; an essential feature of causality is the time lag between cause and effect. Maleness is conferred before baldness, and may or may not be thought of as a cause, but there is no sense in which baldness causes maleness.

Question 4.3

(a), (c) and (d) increase; (b) and (e) decrease.
The catalytic converter is impaired by the leaded petrol, so the three classes of compound that it is designed to destroy increase. These are (a), (c) and (d). The suppression of the reactions shown in Equations 3.1, 4.4 and 4.5 means that the products formed in these reactions decrease. They include carbon dioxide (Equations 3.1 and 4.4) and nitrogen (Equation 4.5). However, the amounts of carbon dioxide and atmospheric nitrogen in the exhaust stream are already large, so the *percentage* decrease in these two gases is not great.

Answers to activities

Activity 2.1

(a) The point which marks the peak of lead production in the Roman period is a horizontal distance from the vertical axis almost exactly the same as that of the division marked 2000. This means that peak production in Roman times occurred about 2000 years before the present date—in about zero AD.

(b) The point of peak lead production lies at almost exactly the same height as the division marked 10^5 on the vertical axis. This means that peak production in the Roman period was about 100000 tonnes per year (10^5 is one followed by five zeros).

Activity 2.2

(a) Only one of the six bars in Figure 2.7 provides information about the region above $35\,\mu g\,dl^{-1}$; this is the one furthest to the right, marked 35–40. Its height corresponds to a figure of about 4 on the vertical axis, so about 4% of inner Manchester men had blood lead concentrations in the range of $35–40\,\mu g\,dl^{-1}$. This breaches the EEC Directive.

(b) The four bars to the right, marked 20–25, 25–30, 30–35 and 35–40, give information about blood lead levels above $20\,\mu g\,dl^{-1}$. The required percentage is obtained by adding together their individual heights which are roughly 27, 9, 3 and 4% respectively, giving a total of 43%. This does not breach the EEC Directive.

Activity 2.3

(a) The typical blood lead concentrations that you encountered in Chapter 2 suggest that the blood lead concentrations have been printed in $mg\,dl^{-1}$ when they should be in $\mu g\,dl^{-1}$. Although you do not have the necessary information at present, you will see later that the lead concentrations in petrol are also wrong.

(b) The blood concentrations quoted are in the range $10–18\,mg\,dl^{-1}$. As 1 mg is $1000\,\mu g$, these concentrations are $10000–18000\,\mu g\,dl^{-1}$. As deaths have occasionally been observed at the $150\,\mu g\,dl^{-1}$ level, Athenians would clearly be dead if these levels were correct.

Activity 3.1

(a) Table 3.2 suggests that the straight-chain alkanes are the hydrocarbon family with the poorest knock-resistance. On the other hand, aromatics are the most knock-resistant. Thus if the straight-chain alkanes in petrol could be persuaded to undergo a chemical reaction, and change into aromatic hydrocarbons, the improvement in knock-resistance would be dramatic. Consequently, during RON measurement, the compression ratio at which the petrol starts knocking should then be greater.

(b) The percentage of TMP in the heptane–TMP mixture with which the petrol is compared at this compression ratio would have to be increased to match the performance of the improved petrol.

Activity 3.2

In the question, the activity was broken down into three tasks which are each given a paragraph in the following answer:

(i) British 4-star petrol has an octane number of 97. This petrol is obtained from light and heavy gasoline fractions with octane numbers of about 60 (Figure 3.4) by catalytic reforming, and by the addition of a high-octane mixture obtained by catalytic cracking of higher boiling temperature fractions such as gas oil. The major contributions to the high octane number of the final product are therefore the deliberate chemical changes made to the hydrocarbon composition of straight-run petrols during these industrial processes. They lift the octane number to about 90 (Figure 3.4). Only the last 6–8% of the octane number is obtained by lead additives.

(ii) If lead were eliminated, the octane number of petrol might be maintained at the original level by intensifying the refining processes, for example by using more catalytic reforming to give a further boost to the concentration of aromatics.

(iii) If lead were eliminated and no compensating changes were made to the petrol, the compression ratios of car engines could be lowered to eliminate engine knock. This, however, would involve loss of power and increased fuel consumption.

Activity 4.1

The argument in the caption is based upon *ratios*: person (b) contains 500 times as much lead as person (a), and one-quarter that of person (c). However, if one looks at *differences*, person (b) is carrying 499 more units of lead than person (a), and 1 500 less units of lead than person (c). This time, we are much closer to our prehistoric ancestors, than to a victim of lead poisoning. Notice that the picture reinforces the first interpretation rather than the second: presentation is important.

The concentration of lead in person (b) is 500 times that of the prehistoric Peruvian whose bones were said, in the text, to contain lead at a level of $0.1\,\mu g\,g^{-1}$. Thus in Figure 4.6, present-day Americans have a bone-lead level of $50\,\mu g\,g^{-1}$. This is over twice the value quoted for Parisians in the text, and its use enhances the environmentalist case. The lead-poisoned value in Figure 4.6 is $200\,\mu g\,g^{-1}$. By using the Parisian value of $23\,\mu g\,g^{-1}$, a simple bar chart, shown in Figure 4.13, emphasizes the *differences* between the three cases, and gives a very different impression from Figure 4.6. You may have chosen some other diagram which emphasizes differences, and is therefore equally effective.

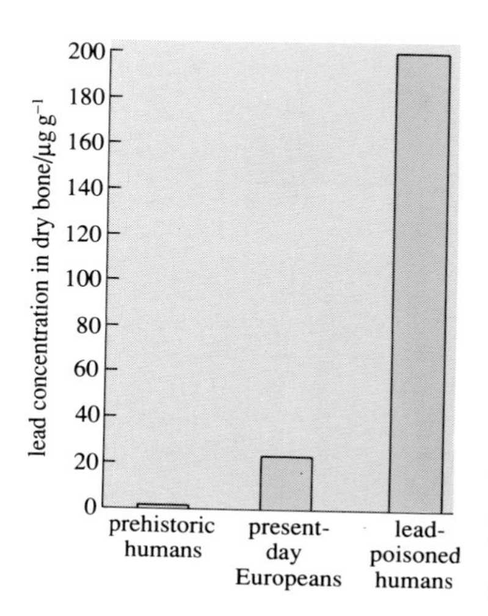

Figure 4.13 Bar chart for answer to Activity 4.1.

Activity 4.2

(a) About 40%; distractibility is dealt with in the first of the 11 questions, and the first set of bars on the far left of Figure 4.7. The high-lead group is group 6, from Table 4.2, and the height of this bar is 40 when measured on the left vertical axis.

(b) About 22%; repeat the operation of part (a) for the lowest-lead group, group 1.

(c) About 29%; from Table 4.2, the children with tooth lead in the range $8.2–11.8\,\mu g\,g^{-1}$ are in group 3. Repeat the operations of part (a) for this group.

(d) 58 and 159 respectively; from Table 4.2, the numbers in the high-lead and $8.2–11.8\,\mu g\,g^{-1}$ groups were 146 and 549 respectively. 40% of 146 is 58, and 29% of 549 is 159. Notice that there were *fewer* children who were easily distractible in the high-lead group even though the percentage was higher because the moderate-lead group has a much higher population.

(e) About 21%—some 79 children; daydreaming is dealt with in question 8, and the eighth set of bars in Figure 4.7. The tooth lead group is group 2, and in the eighth set, bar 2 has a height of 21. From Table 4.2, the population of group 2 is 378, and 21% of 378 is 79.

Activity 4.3

The confounding variables which influence both tallness and baldness are age and sex. By recognizing this, the observed associations can be explained. The association between tallness and baldness becomes weaker in the over-21 group because children are excluded, and children strengthen the association because they are short and hardly ever bald. However, the association persists because women tend to be shorter than men and do not usually go bald. When women are excluded from the over-21 group, the association disappears. The slight reversal is observed because older men are balder and less tall, the latter as a consequence of poorer nourishment in the past and contraction in height with age.

Activity 4.4

The isotopic ratio $^{206}Pb/^{207}Pb$ in both leaded petrol and in airborne lead was about 1.19 in 1974. In petrol it fell to about 1.06 between mid-1977 and mid-1979, a value consistent with the introduction of lead alkyls made from Broken Hill lead with the ratio 1.04. The fall in the ratio for airborne lead was nearly as great: the ratio was down to about 1.07. This suggests that nearly all of the airborne lead in Turin was derived from leaded petrol.

Activity 5.1

This answer should carry a health warning. According to certain scientific or academic ideals, when one presents a case, one should be critical and include a discussion of its weaknesses and uncertainties. By this means, one gets closer to an ideal of shared human truth. I find this admirable, and if that was what I was about here, the answer that follows would include phrases like 'I felt this, but you may think differently because…' But I have made the objective of my answer persuasion rather than truth, and under these circumstances, although one must avoid obvious error, it is best to betray no uncertainty. By comparing your answer with mine, and using each to criticize the other, you can reinstate scientific judgement. To emphasize this, and to provide you with a further comparison, my answer is followed by one given by an Open University student during the pilot study of this book made in 1991.

Answer by book author

The five areas of scientific work in Chapter 4 were the safety margin in blood lead concentration, lead accumulation in the environment, effects of lead on child behaviour and intelligence, the contribution of petrol to blood lead, and catalytic converters. The Royal Commission was most influenced by the safety margin question; this is apparent from the phrase 'We find this disturbing' in the quotation at the end of Section 4.1 which betrays a genuine sense of alarm. Least influential was the effect of lead on child attainment where the quotation is taken up with uncertainties about the conclusions.

In Chapter 5, specific mentions of the five scientific areas only begin with Section 5.3 on the Lawther report. The criticisms made by environmentalists concentrated on the report's sceptical attitude to Needleman's work, and on its choice of a blood lead concentration of $35\,\mu g\,dl^{-1}$ as a trigger for action. In the first case, lead and child attainment is clearly the issue, and in the second case, by implication, it is the safety margin between blood lead concentrations in the general population, and some official threshold of danger. Subsequently, the Turin experiment and other work was used to discredit Lawther's 10% estimate of the contribution of petrol to blood lead, a subject that was also discussed in Rutter's summing up at CLEAR's 1982 symposium. But the high-profile issue was clearly lead and child attainment. This is obvious from the skilful use of the leaked Yellowlees letter to gain media attention, the exploitation of Yule and Landsdown's 1981 study, and the space devoted to the subject in Rutter's summing up. This judgement is resoundingly confirmed by the title of Des Wilson's book.

Thus the environmentalists chose lead and child health to spearhead their campaign, even though it turned out to be the area of scientific work that the Royal Commission found least convincing. This choice, then, had little to do with science. The fate of children arouses intense feeling, and when they are at risk, they have an enormous emotional claim upon the nation's consideration. Moreover, the uncertainty of the effect of lead on child health was no disadvantage because it ensured continual debate and discussion. For both these reasons, the issue was bound to gain much media attention: the environmentalists' choice was scientifically questionable, but politically wise.

Answer by student

(Although this answer consists of more than three paragraphs, the student has broken it down into three parts, (i), (ii) and (iii).)

(i) Section 4.1 refers to blood concentration levels of lead. It is strongly worded against these high levels and is, I consider, the most influential in the fight against lead levels.

Section 4.2.3 chooses to highlight the work of Patterson and Nriagu about the size of anthropogenic emissions of lead. It is quite strongly worded.

Section 4.3.2 deals with the research done to see if there was a causal association between the body burden of lead and childrens' IQ. It is cautiously worded and I would say of little influence, probably the least of the five scientific areas.

Section 4.4 refers to the research on how great a contribution petrol makes to lead uptake in people. It is neutral in tone and the statement that for the majority uptake is less than 20%, seems to detract from the impact of the finding.

Section 4.5.2 acknowledges the fact that unleaded petrol is essential for catalytic converters. It is a statement with an air of resignation to what is virtually now the status quo and as such seems too late to have much influence.

(ii) I think the most important area of the political campaign was the use of the research on behaviour, intelligence and blood lead levels of London children.

'There is a strong likelihood that lead in petrol is permanently reducing the IQ of many of our children', has exactly the right emotion for a political campaign. *Times* readers would be very worried about childrens' IQ, that of their own in particular.

(iii) The contribution of lead in petrol to this high intake was very important to the political campaign. It could be used as a scapegoat to be banished from society and it probably conveniently fitted the hatred many city dwellers were beginning to feel for traffic on their doorsteps.

It is interesting that the Royal Commission came out very strongly against the broad issue of blood lead levels, but that the media and political campaign focused on the fairly narrow and specific issue of what lead was doing to children. This is hardly surprising—children affect our emotions, but broad, balanced judgements on such complex issues are beyond the understanding of most of us and leave us emotionally cold.

Activity 6.1

(a) In some sense, four of the five scientific areas are mentioned in Extract 6.1:

(i) The safety margin in blood lead concentration is discussed in column 2 on p. 73, and emerges most clearly in the sentence, 'Even the present *average* blood lead levels of adults in industrialized countries are not far below those which can lead to overt clinical symptoms of poisoning...'

(ii) Patterson's work on the global accumulation of environmental lead is prominent in Figure 1, and in the first column of p. 73. A good choice of sentence is, 'A recent examination by Japanese and American workers of annual snow strata in N. Greenland has revealed most elegantly that levels of airborne lead have increased markedly since the Industrial Revolution, and very sharply since about 1940...'

(iii) The paper precedes Needleman's 1979 study of a fairly typical group of children, but nevertheless, the risk to children posed by lead is raised in connection with studies of abnormally retarded children in column 1 on p. 72, and at the foot of column 3 on p. 73: 'The authors of ref. 9 reported that nearly half of a group of mentally retarded children had blood lead levels above the maximum level of 0.36 p.p.m. found in a control group of normal children...'

(iv) There is no mention of the percentage contribution of leaded petrol, but the possibility of a substantial contribution is clearly raised at the foot of column 1 on p. 73, and at the top of column 2. This includes a statement corroborated by the Turin experiment: 'Several studies have shown that the principal source of airborne lead in urban environments is the lead-containing aerosol emitted with exhaust fumes from motor vehicles running on petrol containing lead alkyl anti-knock reagents...'

Thus the shape of the 1978–1983 campaign is already prefigured in this 1971 paper. Of the five areas, only catalytic converters are missing.

(b) The *biochemical* effects of lead on, for example, metabolism are referred to right at the start of the paper, and again in column 3 of p. 73. We have not discussed these in the book.

(c) If you are aware that the red colour of blood is caused by haemoglobin, you will have chosen the sentence at the beginning of the paper: 'It inhibits enzyme systems necessary for the formation of haemoglobin through its strong interaction with —SH groups, ...'

(d) To express the figures in $\mu g\, dl^{-1}$, they should all be multiplied by 100. Thus the blood lead levels of the downtown male Philadelphians peak at 0.26 p.p.m., or $26\, \mu g\, dl^{-1}$.

0.26 p.p.m. means that there are 0.26 g of lead in 10^6 g of blood.

10^2 g of blood has a volume of 1 dl.

Therefore 10^6 g of blood occupy $\frac{10^6}{10^2} = 10^4\,\text{dl}$

So lead concentration $= 0.26\,\text{g in } 10^4\,\text{dl} = \frac{0.26\,\text{g}}{10^4\,\text{dl}}$

To change the units from g to μg the required conversion factor is

$$\frac{1\,\mu\text{g}}{10^{-6}\,\text{g}} = 1$$

$$\text{Thus lead concentration} = \frac{0.26\,\text{g}}{10^4\,\text{dl}} \times \frac{1\,\mu\text{g}}{10^{-6}\,\text{g}} = \frac{0.26\,\mu\text{g}}{10^{-2}\,\text{dl}}$$

$$= 26\,\mu\text{g}\,\text{dl}^{-1}$$

Acknowledgements

The Course Team are grateful for the helpful comments and advice received from Dr Trevor Delves, University of Southampton, and Keith Owen, Automotive Fuels Consultant. We also acknowledge the help of the twelve students and five tutors on whom an early draft of this book was tested in 1991.

Grateful acknowledgement is also made to the following sources for permission to reproduce material in this book:

Text

Extract 2.1 'Cleaning Athenian blood', *New Scientist,* **108**, 31 October 1985, IPC Magazines; *Extract 5.1* 'Danger from lead in the air, says professor', *The Times*, 15 February 1971, copyright © Times Newspapers 1971; *Extract 5.2* Dunkley, C. 'Controversy', *The Times*, 31 August 1971, copyright © Times Newspapers Ltd 1971; *Extract 5.3* Vielvoye, R. 'Lead content to be halved', *The Times*, 2 August 1972, copyright © Times Newspapers Ltd 1972; *Extract 5.4* 'Lead poisoning report is "leaked" at road inquiry', *The Times*, 29 November 1978, copyright © Times Newspapers Ltd 1978; *Extract 5.5* 'Lead seen as possible violence cause', *The Times*, 17 December 1974, copyright © Times Newspapers Ltd 1974; *Extract 5.6* Symon, P. 'Government was deceitful lawyer says', *The Times,* 29 November 1978, copyright © Times Newspapers Ltd 1978; *Extract 5.7* Russell Jones, R. 'Lead pollution', *The Times*, 31 March 1980, copyright © R. Russell Jones 1980; *Extract 5.8* Lord Ashby 'Health risks from lead', *The Times*, 2 April 1980, © Lord Ashby 1980; *Extract 5.9* Wright, P. 'New fight on lead in the air', *The Times*, 25 January 1982, copyright © Times Newspapers Ltd 1982; *Extract 5.10* Wilson, D. 'Petrol: must our children be poisoned?', *The Times*, 8 February 1982, copyright © Des Wilson 1982; *Extract 6.1* Bryce-Smith, D. (1971) 'Lead pollution—a growing hazard to public health', *Chemistry in Britain*, **7**, pp. 54–56, The Royal Society of Chemistry.

Figures

Figure 1.1 Science Museum Library; *Figure 1.2* St John's College, Cambridge; *Figure 1.3* Montgomery County Historical Society, Dayton, Ohio; *Figure 2.2* National Gallery of Ireland, Dublin; *Figure 2.6* Moore, M. R. (1986) 'Lead in humans', in Lansdown, R. and Yule, W. (eds), *The Lead Debate*, Croom Helm; *Figure 2.9* Chamberlain, A. C. *et al.* (1978) *Investigations into Lead from Motor Vehicles*, Environmental Medical Science Division, AERE, Harwell; *Figure 3.5* Montgomery County Historical Society, Dayton, Ohio; *Figure 3.8* Cleveland Petroleum Company Limited and United Distillers plc; *Figure 4.4 Geochimica et Cosmochimica Acta*, **45**, p. 1389, copyright © 1981, Pergamon Press plc; *Figure 4.5* Reproduced by permission of Murozumi, M., Chow, T. J. and Patterson, C. C. 'Chemical concentrations of pollutant lead aerosols, terrestrial dusts and sea salts in Greenland and Antarctic snow strata', *Geochimica et Cosmochimica Acta*, **33**, p. 1285, copyright © 1969, Pergamon Press plc; *Figure 4.6* Patterson, C. C. (1983), 'British mega exposures to industrial lead', in Rutter, M. and Jones, R. R. (eds), *Lead versus Health*, copyright © Professor Clair C. Patterson, reproduced by permission of John Wiley & Sons Ltd; *Figure 4.7* Reproduced by permission of *The New England Journal of Medicine*, **300**, p. 691, 1979; *Figure 4.10* Johnson Matthey, Royston, Herts; *Figure 7.1* Reproduced by permission of *Nature*, **353**, pp. 153–156, copyright © 1991 Macmillan Magazines Ltd.

Index

Note: Entries in **bold** are key terms. Page numbers in *italics* refer to figures and tables.